Management for Professionals

More information about this series at http://www.springer.com/series/10101

Nolberto Munier • Eloy Hontoria

Uses and Limitations of the AHP Method

A Non-Mathematical and Rational Analysis

 Springer

Nolberto Munier
Polytechnic University of Valencia
Kingston, ON, Canada

Eloy Hontoria
Campus Muralla del Mar
Universidad Politécnica de Cartagena
Cartagena, Murcia, Spain

ISSN 2192-8096   ISSN 2192-810X   (electronic)
Management for Professionals
ISBN 978-3-030-60394-6   ISBN 978-3-030-60392-2   (eBook)
https://doi.org/10.1007/978-3-030-60392-2

This Springer imprint is published by the registered company Springer Nature Switzerland AG
The registered company address is: Gewerbestrasse 11, 6330 Cham, Switzerland

All methodologies, even the most obvious ones, have their limits.
Paul Feyerabend in *Against Method*

Preface

Objective of This Book

This book is perhaps different or not similar to other technical books written on multi-criteria decision making (MCDM), because of its purpose. It does not describe a new methodology, make a review of existing methods, propose new applications of these methods or introduce new techniques.

The objective of this book is to make practitioners aware of the areas where the analytic hierarchy process (AHP) method can be advantageously used, as well as its structural inability for handling complex scenarios.

All MCDM has limitations when considering modelling and solving complex problems. The AHP is not an exception, and its use in complex scenarios is considered impracticable, because it has not been designed to tackle complexity.

The reason why this book addresses only the AHP is because it is by far the most utilized method, with hundreds of papers related to its large variety of problems and fields. However, examining those publications and analysing the problem, it is quite evident that many practitioners are not aware of this limitation, and they assume that the method can manage complex scenarios.

In addition, there is evidence of an excessive relaxation of certain characteristics of the scenario and in adjusting reality to the method. As a simple example, consider that very often AHP is used when there is blatant evidence, even at first sight, of existing relationships among criteria, something that the method does not accept. This fact is most of the time ignored, or, as these consultants were told by a practitioner, in relation to a paper with a real project that she and her colleagues wrote, 'Well, yes, we noticed that, but decided that it was not important'.

Which Is the Best MCDM Method?

This is one of the most frequent queries from practitioners.

No methods or processes are recommended as a better choice; MCDM involves many different types of problems that normally need to be solved by specific methods. After a prior selection, a method is chosen, because a practitioner knows it or owns the software or because it has been used in a similar scenario. This is not an acceptable practice. Most methods have been designed to address a specific type of problem, and they can't be used to solve all different kinds, let alone complex ones. Consequently, a practitioner, after thoroughly examining the problem, must select the method that best fits it. To that effect, we present Table A1 in the Appendix that proposes an innovative table for selecting the appropriate MCDM method for a given scenario, complemented by an example in Table A2.

Which Is the Most Used Method?

Statistics involving data collected over decades show that the most used method worldwide is AHP. Probably the reasons are that it is easy to understand, easy to learn and not mathematically complex and has a good software, but most importantly, from the authors' point of view, it gives users the feeling that they can solve complex problems based on their own perceptions. It is precisely the objective of this book, since the authors do not agree with that notion they consider misleading.

Which Are the Most Convenient Scenarios for AHP Application?

AHP has been designed to solve problems where the direct and personal involvement of the DM is necessary and dominant, that is, problems where the assessment of the importance of a criterion, an alternative or a situation is based on individual preferences, tastes, needs and intuition. This further means that it is applied to problems where personal considerations are a priority and which often take precedence over objective and definite features. The procedure is a way of measuring how the DM sees a problem, instead of considering how in reality it seems to be, especially in uncertain situations.

As its creator (Saaty 2008) asserts, the 'AHP is a theory of measurements through pair-wise comparisons', and it measures intangibles in relative terms and based on psychological assumptions. It appears not to be a coincidence that Saaty developed this method when working with the military and, possibly, as an answer to their needs.

However, the authors believe that this method of addressing scenarios can't be applied to most real-world problems in industry, commerce, construction, transportation, location, manufacturing, etc., where there is no room for personal judgements. The result of a problem must reflect that existing facts can't be ignored and, if possible, quantified with reliable and objective data, for which AHP was replaced with DM preferences.

Thus, the immediate conclusion from these reflections is that in general, two different kinds of scenarios exist: those where the personal involvement of the DMs is essential and those where real values, circumstances and situations are vital and must be forcefully considered.

This does not necessarily mean that a clear separation between the two different scenarios exists. Nowadays, it is practically impossible to find a real-world problem where there are only objective features, for the simple fact that all projects, in one way or another, are related to and developed for the society, as a consumer of goods and services and as the main modifier of the environment. Consequently, the subjective opinion of potentially affected people must be considered.

It does not imply that the AHP technique has to be included, since these subjectivities can be addressed in different manners, for instance, with surveys and polls, and further aggregated through statistics, where there is no need for pair-wise comparisons.

The opposite is also true, i.e. in the first case of scenarios analysed, there could be, and normally are, quantitative aspects that have to be considered; however, AHP does not take them into account or reduce them to a pair-wise comparison.

As a conclusion, the authors infer, in view of the fact that the core of the AHP method is strongly related to a personal type of scenarios, that it is not up to be employed for complex problems. This is not however its only limitation; in this book, it is demonstrated that there are some more, not only based on the author's opinion but also considering the judgement of more than 100 reputed researchers.

What Is a Complex Problem?

According to Singh (2016), 'The project is complex enough that subtasks require proper coordination and control in terms of timing precedence, cost and performance'.

Regarding MCDM, we expand this definition as follows:

Complex projects are those that, in addition to having the typical scheme of several objectives, as well as alternatives subject to a set of criteria, include many specific characteristics, demands and conditions, alone or combined, inherent to a problem, which can also be associated simultaneously to various scenarios and in different time periods.

Which Other Features Impede AHP to Address Complex Problems?

The authors understand that there are three reasons for this inability.

First
Complex projects are normally overwhelmingly based on objective data and work with man-hours, investments in a certain currency, percentages, resource availability, quantified low and high limits, algebraic formulas, volumes, etc., that is, with tangibles, and often mixed with subjective or intangibles such as population needs, people opinion, perceived advantages, erosion, preferences, externalities, etc., where quantities and units of measure are probably absent and most of the time uncertain. These subjective criteria are a genuine necessity, because whatever the project is, it can't ignore the society it will serve.

However, normally, in real-world scenarios, usually complex, these subjective and psychological aspects constitute a tiny percentage of the total criteria needed for alternative evaluation, while their participation is large and sometimes exclusive, in other types of scenarios, such as selecting a place for a holyday, opting for a university to enrol or choosing a house to buy.

In individual and corporate projects, there is personal involvement as well as a personal interest in the result, while in complex projects, the DMs are just making their opinions known, contributing their experience in analysing the result and in making recommendations, and most probably, the outcome of the project will not have any influence in their lives.

Second

Pair-wise comparisons are good and useful for analysing subjective issues, but not for objective ones, at least in the MCDM setting.

Their creator (Thurstone 1927) defined them very precisely as 'THIS is an attempt to apply the ideas of psychophysical measurement in the field of social values'.

Of course, as said, psychological-based opinions may be necessary in subjective criteria in real-world scenarios, but it does not mean that we have to work with pair-wise comparisons. It is possible to analyse each criterion independently and utilize a 1 to 10 scale to measure it, or a Likert scale, and as long as we use the same scale for all criteria its type does not matter.

Therefore, why do we compare two criteria when we only need the value for one? Observe that in appraising each criterion independently we are determining its absolute importance, not trade-offs as in AHP.

There is a contradiction in the AHP approach using pair-wise comparisons, because it assumes that criteria are independent. Hence, why do we value one as a function of the other, even regarding the same objective?

Consider projects or options. If two diesel generators have reliable and documented efficiency values for each machine, duly certified by the respective manufacturer and on the same base, what is the logic in making comparison between them? If machine A has a thermal efficiency of 48.6 % and machine B has 49.1 %, obviously, B is better than A, in terms of efficiency, but it may be the opposite when maintenance costs are considered.

Third

AHP is based on a hierarchical structure, that is, a lineal or military organization with a top-down flow of commands and preferences and where decisions from upper levels are rarely discussed. But normally, complex projects don't follow this structure, because there are many relationships between the diverse components and multiple levels, including horizontal links as well as correlation and feedbacks, and usually there are different opinions that are considered and discussed, even those coming from lower echelons.

The hierarchical structure is, in the opinion of the authors, the main hurdle of AHP. Most probably, Saaty understood this situation and wisely developed his 'analytical network process' (ANP), using a network structure instead of a hierarchical one, which makes more sense. It gives more latitude by incorporating some characteristics from a project, for instance, precedence; however, this is not enough, and ANP is still very limited for complex scenarios.

Here there could also be combined inclusion and exclusion, as well as certain relationships among criteria and demands, and/or relationships between alternatives and criteria, that are impossible to consider, unless a binary matrix is used.

The impediment in the hierarchical structure is that in many cases, in addressing complex scenarios, it can't incorporate the particularities that each project has. This is the main reason, by which at the authors' judgement, AHP is not adequate to address complex scenarios, simply because its structure is not prepared for that.

Its hierarchical structure is however useful for examining and analysing some relationships in a complicated problem, but not for its solving. This assertion is not a presumption or suspicion; it is duly justified in this book by reasoning and using many examples. Researchers such as Öztürk (2006), Song et al (2016), Belton et al (2010) and Chung et al (2005) also find it not appropriate.

Where Can AHP Be Advantageously Used?

AHP can be satisfactorily used in many different areas such as:
- Individual and corporate scenarios
- In the military
- In personnel selection in organizations
- In selecting stocks to purchase
- In universities, where the opinions and experience of professors and students' feedback are important
- In health-care issues, related for instance to different surgery procedures, where the opinion of different doctors must contemplate also the wishes of the patients
- In daily activities such as purchasing a house, selecting a restaurant or a movie, deciding a place for holidaying, etc., and of course, for the person taking the decision, who may be indifferent to others

These personal scenarios may be difficult, painful, hurting or painless, but not complex.

But What Is the Main Difference Between Trivial and Complex Problems?

All personal or corporate projects, with the possible exception of the military, have a common denominator: the consequences, favourable or unfavourable of the decision, fall on the person, corporation or entity making the decision, either by a sole decision maker (DM) or by a group of DMs or by the stakeholders. That is, consequences are limited to the people that take the credit for a successful decision or the blame for a bad one. This is a normal occurrence in our lives, as we constantly make decisions that bring joy or sadness, happiness or sorrow, satisfaction or gloom.

They are normally identified as 'trivial' or 'naïve' or 'straightforward' problems. However, don't make the mistake of belittling them by thinking they are easy to solve, unimportant or without complications. All projects are important, and their difference is only based on the number of special characteristics and their relationships. They can be difficult, laborious and intricate; therefore, these terms simply express that they are not complex.

Every problem is important to the interested party but not all problems have the same importance or affect the same party.

Hiring or expanding an enterprise is important for the company and perhaps for its employees. Building a hydro-electric dam is important for the society as a whole, because it can improve the quality of life of thousands or millions of people, reduce the country import of oil and generate revenues for tourism in the lake behind the dam. It is obvious that the two scenarios have very distinctive scope and can't be treated in the same way.

Problems such as deciding about a location for an industry, determining a set of environment indicators, establishing a transportation urban network and supply chain, selecting the best policies for urban wastes, choosing a route for an overhead power line, developing a composite indicator, etc., are complex undertakings. This is due to their existing multiple relationships but also due to the unique characteristics of each scenario, such as correlation, dependency, inclusive and exclusive alternatives, joint ventures, etc.

They are also complex since in different ways they may produce changes in the life of millions of people, and each one may be related to a different undertaking, such as a nested project, with distinct characteristics, considering for instance social and legal aspects. Examples abound, such as the construction of the Three Gorges Dam in China that involved the mandatory relocation of hundreds of thousands of people, a huge undertaking by itself.

Is it then rational to think that the same methodology can be applied in hiring people and in analysing a railway network?

It is rather obvious that DM responsibilities largely exceed those regarding trivial problems and that no personal interest is acceptable. However, in AHP these aspects are ignored, and many times, it is assumed that the DM may decide in lieu of millions of people, which in addition contradicts Arrow's impossibility theorem (Arrow 1951).

There is another issue that pertains to the size of the problem; in complex projects, we are talking of maybe a small number of alternatives but perhaps with hundreds or thousands of criteria. This can be seen in a river basin analysis scenario. The necessity of performing thousands of pair-wise comparisons precludes the use of the AHP, with the potential perspective that the work has to be partially redone if, as usually happens, even a sole alternative is deleted or added.

If There Are No Pair-Wise Comparisons and Preferences, Which Is Then the Role of the Decision-Makers, or Are They Unnecessary?

DMs are the most important component in a MCDM problem; they are those who verify with the corresponding sources that the alternatives are feasible and select the criteria and the MCDM method to be used, and they are those who examine the final result given using the mathematical model and decide if the solution is feasible and robust; therefore, the action of the DM is crucial in this endeavour.

If there are subjective criteria, they must design the questions for people to answer and translate linguistic responses into cardinal values. Their judgement must be in line with that of the consulting people and they must adhere to their decisions, in thoroughly examining the characteristics of each project, in working with 'What IF....' scenarios and in analysing the potential political future problems and consequences, as happens in real projects, such as the construction of a railway from Eastern China to Western Europe.

Is There Any Academic or Scientific Support to the Authors' Assertion Regarding the Incapability of AHP to Handle Complex Problems?

Based on numerous works from specialists for decades, there is evidence that this method is mathematically flawed. However, this book does not address this matter, and consequently, it does not examine it from the mathematical point of view.

The authors support the idea that AHP can't handle complex problems. However, it is not a simple belief, idea or conjecture; the assertion is grounded on 30 different related issues, in using logic and reasoning, and illustrated using real-world examples. It reproduces verbatim the opinions and comments of 105 reputed researchers, including of course those from Saaty and supporters. Out of that number, 101 researchers give evidence that backs up what this book upholds. We leave the readers to extract their own conclusions.

This book develops as follows:

(a) Introducing and explaining the reasons from the authors' point of view.
(b) Analysing the characteristics of the hierarchical structure, which is the core of the AHP and its inability to consider situations other than those that can be inserted into this type of structure, and making it not adequate to portray and model complex projects.
(c) It challenges the common assertion that AHP can solve complex problems, which seems to be a set phrase, since it frequently appears in technical papers, but that nobody, in the authors' knowledge, has considered the necessity of demonstrating its veracity.

In order to discuss the accuracy or mendacity of this assertion, one needs to define first what implicates the concept of complex problems, and this is the next step. However, instead of dwelling in philosophical discussions, it is better to use some complex scenarios that are common in the real world. If the method can solve them, we can conclude that the statement is correct, and the matter is settled; if not, obviously the statement is false. Of course, only one case is not enough, and for that reason ten different cases are proposed.

In that sense, this book puts forward a series of real complicated and complex scenarios, in different fields such as: city infrastructure, housing development, location selection, portfolio of construction projects, hydro-electricity, regional structure, access to airports, environment, urban structure and oil refinery (production, storage and transport). Not a single case can be solved by AHP.

With these examples, the reader can examine these proposed scenarios, which vary in complexity, and decide if any of them can be solved by AHP. It will be up to his/her judgement, which may be much more valuable than the authors' opinion.

If the readers conclude that according to their own assessment these scenarios cannot be solved by AHP, that is enough proof to invalidate the claims about their ability in solving complex scenarios.

The authors believe that it is preposterous and ludicrous to consider that complex situations can be modelled by a hierarchy, doesn't matter how complicated it might be.

This is the core of this book, since it shows the inability of AHP to model them, as well as the misleading assumptions to justify its process.

(d) Examining the rationality of AHP method and the reasons for this critique.

(e) Making an inventory of shortcomings. This is probably the most important part of this book. In here, 30 shortcomings, drawbacks, limitations, inadequacies and deficiencies are analysed, using only examples, reasoning, thinking and common sense and a language that everyone can understand.

These 30 subjects are examined, including definitions, demonstrations and comments from the authors. Following this examination and analysis, each subject followed a procedure that, as far as the authors are concerned, is novel; the book contains a list of published opinions, comments and judgements on each subject, from a total of more than 100 researchers; these expressions are reproduced verbatim, naturally, citing the author and the source. Since sometimes these concepts are inserted in a long text, the authors have put between brackets what the respective author is referring to; other than that, there is no intervention or opinion from the authors.

These judgements and views, coming from so many researchers, are considered the most valuable part of this book, for they constitute an ordered collection of judgements, positives and negatives for each one of the 30 subjects, except in 3 of them, because the authors couldn't find the researchers' opinions on them. They are independent of the authors' comments on the same subject and then can be considered completely unbiased and neutral regarding our opinions, since they are not used to support or deny them.

In addition, no post judgement is made by the authors regarding the agreement or not from the cited researchers and their own comments.

To facilitate the comparison of the agreements or disagreements between the authors' opinions and those from other colleagues, each subject is followed by a table, with both vis-à-vis opinions.

The number of researchers addressing a particular subject, which indicates the interest on each one, is shown in Table 9 (only the first author in each case). There are 105 researchers cited and their opinions are reproduced; out of those, 89 (96 %) researchers support the authors, and 4 (4 %) think the opposite, the latter exclusively coming from the three researchers: Thomas Saaty, Luis Vargas and Patrick Harker, who happen to be Saaty's co-authors.

That is, with the exception of the author of the method and his two colleagues, no researcher comes to justify it.

(f) Including a last chapter where a framework in MCDM that deals with complex projects is proposed.

(g) A glossary is included for readers' convenience.

(h) Including a table in the Appendix as a tool to support the DMs for selecting the most appropriate MCDM method to solve a problem.

Kingston, ON, Canada   Nolberto Munier

Cartagena, Murcia, Spain   Eloy Hontoria

Acknowledgements

Eloy Hontoria is grateful to the Ministry of Science, Innovation and Universities (MICINN, Project RTI2018-099139-B-C21) and FEDER for their support.

Contents

General Concepts

1

A well-formulated question usually holds the answer.
Anonymous

Abstract

This book consists of a series of discussions and analysis on certain issues identified as 'subjects', regarding the Analytical Hierarchy Process (AHP) with respect to its structure and shortcomings. It is addressed to Multi-Criteria Decision-Making (MCDM) practitioners, professors and students in the process of learning or using this method for decision-making.

Keywords

AHP · MCDM · Real-world scenarios · Shortcomings · Drawbacks · Common sense

1.1 Introduction

This book consists of a series of discussions and analysis on certain issues identified as 'subjects', regarding the Analytical Hierarchy Process (AHP) with respect to its structure and shortcomings. It is addressed to Multi-Criteria Decision-Making (MCDM) practitioners, professors and students in the process of learning or using this method for decision-making.

It is not aimed to mathematicians and then it does not utilize mathematical language involving demonstrations, formulas or axioms; it uses instead plain English that everybody can understand, irrelevant to the field of activity, technical preparation and expertise. It is performed using two boundless resources available to anyone: reasonableness and common sense.

In so doing, the book tries to answer many questions from practitioners, that is, queries that users have been formulating since the 1980s, when the method was introduced as an MCDM discipline, and that have received no rational responses or explanations.

These authors have been analysing different MCDM methods, and most especially, AHP, from the point of view of their rationality and their utilization in MCDM scenarios. Unfortunately, they found by far more cons than pros and believe that it is necessary that users be told about the drawbacks and inadequate structure of the AHP.

These authors criticize the Analytical Hierarchy Process (AHP), with no intention to belittle it but to show its drawbacks, since they interpret that the users have the right to know about the capabilities of the method they are using or plan to use.

There are possibly hundreds of papers dealing with the mathematics of AHP. As mentioned, the authors are not going to add a new one to the already existing list, although we believe, as other colleagues, that this method is mathematically flawed. Only our personal experiences and examples are used here; they reflect real-world scenarios, and possibly they are, for many practitioners, more understandable than mathematics.

AHP is by far the most used MCDM method and with hundreds, if not thousands, of papers published that illustrate its use on different applications. This book aims at making users aware about the characteristics of this method, its merits as well as shortcomings and weaknesses, and in spite of its popularity, its inability to solve real-world problems, not even those that are just a bit complicated, let alone to deal with complex scenarios. The reason is simple: the method has not been designed to tackle serious problems, only trivial ones.

However, an immediate and logical question arises: If the method has so many drawbacks, why is it used by so many people and in many different scenarios?

The answer is straightforward: it is attractive, is easy to use, has a friendly software, and because it gives the decision-maker (DM) the feeling that he/she is really solving a complex problem based on his/her preferences, and in addition, due to the fact that many users, even AHP experts, believe or want to believe that the method can be applied to any scenario. In an occasion, one of the authors (N. Munier) was told, in writing, by an experienced AHP consultant, that he recommends his clients to use AHP, doesn't matter the type of problem.

However, the users should ask themselves or wonder how it is possible that complex scenarios, involving engineering, economics, environment, social issues, complex interrelationships between alternatives and criteria and so forth can be addressed so easily, simply according to their preferences.

It is difficult to think that an architect, for instance, may decide to erect a large building using either wood, steel or concrete beams, based on his/her personal preferences on which material to use. This lack of common sense is the main drawback of this method, to say nothing of using a scale to determine how many times the concrete is superior to steel, something that not even engineering technical studies can assert because it depends on its intended use.

Take for instance a manufacturing scenario where the alternatives are subject to several criteria, two of them being *Production cost* and *Environmental contamination*. Is it acceptable for the DMs to establish their preferences saying, for instance, that the cost is three times more important than environmental contamination, regarding the main objective, *Maximize benefits?*

No, because in so doing, the DMs are ignoring, among others, how much could be the cost in ancillary equipment to decrease the contamination to acceptable limits. It could be that this cost is higher than the production cost. They can argue that the cost of this ancillary equipment was already considered when the production cost was computed. However, that would be unreal, because the production cost is a direct cost, which totally varies with the number of goods produced, whereas the cost of ancillary equipment is a fixed cost and thus is not related with production.

Let's go now to more subjective comparisons. Assume that, in regard to whatever main objective, the DM needs to compare two fruits such as *oranges* and *pomegranates*, and he decides, based on his taste preferences, that oranges are much tastier than pomegranates, and for that, the Saaty scale gives a score of 5. He has the right to establish his preferences – which by the way he may change – however, how is it possible to give a crisp number to something so difficult to evaluate as taste, which, in addition, can change depending on many different things such as water content, sugar, minerals, origin of the fruit and so on?

These very simple examples show a crucial aspect of the AHP. It works on assumptions and theories without considering practical aspects. Consequently, practitioners, when working with real-world problems, are at a loss because they don't know – and with reason – how to proceed if the project shows a particular feature, and in that case, the DM just decides to ignore it.

In these authors' opinion, AHP could be useful to find a partial solution to trivial problems where the people affected by the consequences of a decision are themselves: for instance, the typical example of purchasing a car, renting an apartment, selecting a moving, choosing a medical treatment, deciding on a restaurant or selecting people to hire for a company. It is said 'a partial solution' because even in trivial problems AHP is grounded on dubious procedures.

It goes without saying that decision-making is an activity that needs a very important dosage of subjectivity, and nobody can deny it. Therefore, these authors are not assuming that MCDM is a pure algorithm contraption; it is useless without the knowledge, expertise and reasoning of the DM. The trick, in our opinion, is the time when the DM intervenes.

One of these awkward features for some AHP users is to realize, by simply examining the criteria on a certain scenario, that they are related, when the method considers them independent.

This means that the AHP method in this circumstance can't be utilized, however, the DM decides to use it anyway, since he considers that it is not a serious problem and, in addition, not important.

1.2   Conclusion of This Chapter

This first chapter is an introduction to the main subject of the book, that is, to demonstrate the limitation of AHP to solve complex scenarios (described in Sect. 2.1). As mentioned, it is not written in mathematic language, and it aims at pointing out the deficiencies and drawbacks of the method.

Consequently, it may be considered a technical book because of the subject it addresses, but it is in fact a publication to alert practitioners regarding the scope, utility and reliability of this method.

It uses reasoning and especially real-life examples that show the weakness, flaws and shortcomings of AHP. They are not only the opinion of the authors, but also those of many scholars. Authors, as human beings, may be mistaken, confused or without enough knowledge of this subject, but it would be very difficult to believe that more than 100 scholars supporting their ideas in one way or the other are also wrong.

The book aims at being a source for practitioners to consult if they decide using AHP, by alerting them about what is not so noticeable or visible in a method which appears to be able to solve many types of projects.

The Hierarchical Structure

2

*The errors of a theory are rarely found in what it asserts
explicitly; they hide in what it ignores or tacitly assumes*
Daniel Kahneman

Abstract

This chapter addresses a fundamental issue in Multi-Criteria Decision-Making
(MCDM) problems consisting in how to model complex scenarios. It starts by
defining complex scenarios and examines the hierarchical structure followed by
the AHP method, considered not suitable for modelling them, as well as their
relationship with the decision-maker. It provides a hierarchical structure where it
can be seen the impossibility in modelling certain characteristics, that may be
fundamental in a scenario.

It also discusses Saaty's concept of impacts and feedbacks. The first is impossible to in structure, while the second is inexistent in MCDM scenarios.

Keywords

AHP · Complex scenarios · Modelling structures · Alternatives relationships ·
Impact · Feedback

2.1 What Is a Complex Scenario?

It is not easy to define what a complex scenario is, since many practitioners refer to
it, but without a clear understanding of what it really means, scope-wise, and in
addition, there is not a definition but several.

For some researchers, a complex scenario exists when there are many criteria, numerous sub-criteria, and interrelated, as well as clusters, and many alternatives and sub alternatives, and this is correct, only that this definition is too basic. It is necessary to have this issue clearly established to be able to gauge if the MCDM methods are really dealing with them. Nowadays, there are not simple projects or scenarios, because in general all projects demand conditions, establish settings are complex environments, and even incorporate supplementary projects, not related with the main objective, that were not even hinted at the beginning of the analysis and which are appearing along with the examination of the scenario.

That is the case of projects in certain parts of the world, where in addition to the complexity of the project itself, it is necessary to consider isolation, existent diseases, aggressive inhabitants, political rivalries and so forth. As an example, the Panamá Canal was started by France in 1881, and the work was further abandoned because of the great rate of mortality among workers, due to malaria. Apparently, this threat was not foreseen or its danger underestimated in the initial plans.

When in 1904 the United States decided to continue the project, they added some complementary projects that were not directly related with the excavation initiated by the French, but that had an enormous importance in the scenario. They were diverse undertakings to get rid of the mosquito-borne infectious disease by fumigation and other measures, but what was important in decision-making is that those projects should precede the main excavation project. In this way, precedence was incorporated into the main project.

Until the mid-twentieth century, the main and unique objective of projects was to maximize benefits or minimize costs, and traditional accounting procedures were employed, such as the Internal Rate of Return (IRR), the Net Present Value (NPV), the Benefit/Cost ratio or the Payback Period. These were simple scenarios and with only one objective: to make money or to reduce costs, without consideration of related advantages or disadvantages.

That is, the objective was always economic, and thus, not considering how projects influence society, impulse development, alter the environment, consume resources, transform landscapes, modify transportation, and very often affect the way of life of people and wildlife.

However, all changed when analysts realized that economics could no longer be the only objective; that society needed to participate in projects decisions and that exerted pressure to partake in decision-making, demanding that new factors should be considered, as those related to workers, safety, contamination, loss of natural capital, risks, health and so forth, that is, the human factor, the environment and sustainability. Consequently, the introduction of these concepts indicated that old methods were not prepared to deal with these new inputs.

This was the starting point for scientists and scholars to develop methods being able to find solutions that reflected the new concepts, and researchers like Kenney and Raifa (1976) began with the concept of utility, or satisfaction rendered for something done, and opening the gate for looking means to get it. One very important concept that appeared was that a certain objective, be it maximize benefits, minimize costs or equalize a goal, ought to be dependent to the compliance of a series of conditions, not only one. These conditions were called *criteria*.

Since then, projects and scenarios were subject to the compliance of criteria that could refer to aspects apparently without connection with the project, such as the environment, the wildlife, the loss of natural capital or the fate to those people who would be affected by the project.

However, which complicated the issue the most, was the fact that every project had to comply simultaneously with all the criteria to a greater or a lesser degree. This is what is known as *multi criteria*, and naturally, from there came this process to be known as *Multi Criteria Decision Making* or MCDM.

Therefore, projects were by far more complicated because of new demands, different units of measure, and with the need to consider quantitative and qualitative criteria, and of course with increasing uncertainty, which prompted Zadeh (1965) to develop a system to treat uncertainty more rigorously, the fuzzy approach. However, researchers realized that it was not very realistic to work with only one objective, and this was very significant; consequently, the mono economic objective of the first models was replaced by many different objectives, subject to many different criteria.

Normally, it is impossible to comply with all objectives, because the increase of one of them probably will produce the decrease of another/s, as is the case in conflicting objectives, as for instance asking at the same time to minimize contamination and maximize electrical generation with fossil fuels. Because of this impossibility, the solution should be a compromise solution (Zeleny 1974), and this is the concept that provoked the searching of many methods.

That is, the procedure is not to look for optimal solutions – which are normally inexistent – but for feasible and convenient ones, and then, using *heuristic algorithms* instead of optimization ones.

These seminal ideas and concepts prompted researchers to find new methods that considered the new conditions. Probably, the first method was the Weighted Sum Model (Fishburn 1967), while Roy (1968) developed a method based on outranking, called ELECTRE, followed by Brans and Vincke (1985) with a method called PROMETHEE, and further by Saaty, with his AHP (1986) and ANP (1994). Another important method was TOPSIS (Hwang and Yoon 1981), based on the distance to an ideal point.

2.1.1 The Modelling Structures

It is true that in all organizations, there must be some command chain from the top level down to the inferior levels, mainly based on responsibility, importance of each level and so forth, but there are also other types of organizations that contemplate transversal relationships, and something similar is needed if we want to model a complex scenario. This is probably the main drawback in AHP; it is too simplistic, and notice that other MCDM methods do not follow it.

In MCDM methods, the DM is indeed at a high level and subject to a chain of command mainly from the owners and stakeholders, that is, it is a top-down approach, from general to particular, but it does not mean that in modelling a problem, the DM must adjust to that scheme as AHP does. It establishes one objective – which is indeed strange enough, since normally there are several – and followed by criteria, criteria clusters and sub-criteria, and at the very bottom, the alternatives. It is believed that it would be more logical to start with one or several main objectives, determine the alternatives or options to attain them and finally establish which are the conditions or criteria that the alternatives must be subject to.

This is the process we normally follow in our personal life. If we decide to take a holiday and the objective is to enjoy it at the lowest price and trouble, then it is natural to first consider the potential sites; this is typically what everybody does. If our alternatives are the beach, the mountain or a cruise, then we need to know what each destination offers, such as exciting social life and entertainment, or a quiet and relaxed scenario, or the opportunity to discover new scenarios, or to taste exotic dishes and discover other cultures and so forth. It is essential for us to know about costs, accommodation facilities, risks and so forth, and this is the data that we need to take a decision.

Lets' see this from another perspective. We can say that it does not matter the place, since we do not have too much money and want to spend it wisely, and for us, price is more important than comfort.

In so doing, it is assumed that is always true to that same amount of money can buy a meal, a trip or a hotel, irrelevant of the site, and of course, that is not true, since some of these commodities have different prices and comfort, depending on many different things, and therefore, it could be that in one site there are comfortable hotels, but less expensive than in others. That is, if we don't know how each different destination cost is, what is the advantage of knowing how much money we have?

Something similar happens in real-world projects. A company may have a five-year plan to develop and market a series of products, but first, it must know which these products are. How else can the company estimate market share, transportation, market penetration, manpower needed and so on? Just by preferences?

As another typical example very often published, assume a fresh high school graduate who wants to be a geologist. He/she must first learn which are the universities, that is, the different institutions that impart that discipline, and after that, he/she can study their prerequisites and conditions such as tuition fees, possibility to develop valuable relationships for the future, university quality level, prestige and so on.

From these elemental analyses, it appears that it is incorrect to apply hierarchies and to decide conditions without knowing the options.

In opinion of these authors, the AHP weakest point is that it was theoretically conceived based on a psychological theory and elemental mathematics and ignoring what really happens in the real-life; it overlooks the practical point of view of projects and their complications and variations. All is reduced to model a problem according to a hierarchy, not realizing a fundamental issue: every project is different, has its own limitations, its own needs, its own demands and, accordingly, needs to be tackled differently.

Does it mean that each project needs determined software? Yes, it could, or as an option, use a method with enough flexibility to adapt to different scenarios configuration.

2.2 Complex Scenarios and AHP

A very usual comment about the AHP, and mostly stated by authors describing the method, is that it can solve complex scenarios. It is an incorrect statement, because AHP is based on modelling a problem through a hierarchical structure alone, which is not adequate for complex scenarios. It could be useful for some simple scenarios, but inadequate for most others.

The reason is that sometimes it is impossible to model from the hierarchical structure, because it does not allow for registering many aspects found in complex scenarios; one example is the above-mentioned Panamá Canal and the need to develop projects following site conditions.

The hierarchical structure is without a doubt very useful to analyse relationships among the components of a project; the best example probably is the *Work Breakdown Structure* (WBS), which is a graphical hierarchical model of something to be manufactured or executed, as a product or a project, and as its name hints, breaks a complex construct in parts for its analysis. It is a very convenient and extensively used device, where diverse elements are linear and transversely connected to give inputs to another at a higher level; however, it can't be solved in parts, but as a whole, as a system.

These connections are something that AHP does not share, because it does not recognize, as ANP, transversal relationships such as between criteria and among alternatives. It resembles more to the old and known lineal organization or line structure, also branded as a military structure, where there are different levels with a vertical chain of command. The AHP hierarchical structure follows the same pattern. Just find in the Internet the disadvantages of this rigid structure that was replaced by more rational and especially practical relationships, like the matrix organization, because it showed that it couldn't cope with modern business and manufacturing practices, and the latter replicates in AHP, since it can't handle scenarios in which components are related.

Figure 2.1 represents a rather complicated structure, with a main element, three criteria and many sub-criteria. It is a modified hierarchical structure, because it admits transversal relationships, as well as certain conditions that cannot be modelled in a conventional hierarchical structure.

The example refers to the structure for a country at the top and two main cities, which in turn comprises different districts within each city. Each district has a portfolio of projects of alternatives, and some of them have sub-projects. It is a good and useful graphical representation of the country, cities and alternatives, and that is the most that this type of structure can offer.

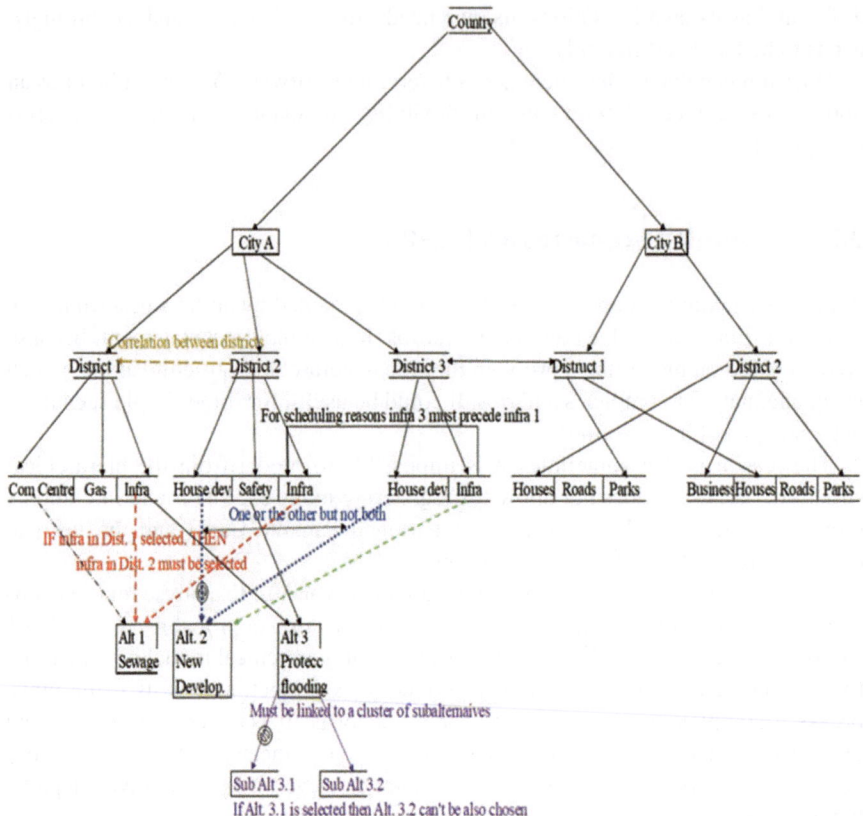

Fig. 2.1 Scheme of a complicated structure

What we want is to determine a ranking of alternatives considering all the conditions, other than the typical criteria, and which are expressed in colours.

Thus, in red, we indicate *dependency* and a mandatory *restriction*. A project for sewerage construction must involve infrastructure in district 1 and the same in district 2, but not in district 3, and in each case, up to different budgets.

In blue, it is established that if one alternative is selected, another can't be selected, that is, *exclusiveness*. There is a project involved with house development that concerns only districts 2 and 3. However, they are exclusive, that is, only one of them can be executed, may be because of funds shortage.

In purple, we indicate that alternative 3 has a *cluster* with sub alternatives, and again, it is one or the other.

In dark yellow, we inform the software that there is *correlation or impact* between district 1 and district 2, for instance, a house development on district 2 may produce that people in district 1 are no longer with a clear view of the mountains.

In black, we indicate that infrastructure in district 1 has to precede infrastructure in distinct 1 must precede alternative 3. For instance, to prevent landslides damaging in district 2, it is necessary to reforest areas in district 3. That is, *conditioning*.

Of course, there are more conditions that may be considered.

We have indicated all these restrictions in the diagram, but *they are not a model. A model must be built as a matrix* that contains all the restrictions shown in the diagram.

Therefore, we have to find a way to model them. That is, we can use the modified hierarchical diagram to get an idea of the whole problem and, from there, to model this scenario in matrix format.

What is needed is a framework that could be built within a hierarchical structure but that allows incorporating what the scenario demands, and this can be achieved with a framework consisting of, at least, a pair of different matrices, the normal one, with the performance values and another one representing conditional and comparative scenario demands, normally, in a binary matrix.

Saaty, after developing the AHP, produced the Analytical Network Process, or ANP, where he uses not a hierarchy but a network, which makes more sense, since in ANP one can consider interrelations, among all elements of the scenario. However, these interrelations are reduced to simple links between the diverse elements, and it is not possible, for instance, to register the impacts on one element over another or model any of the conditions commented above.

2.2.1 The Influence, Impact, Feedback and Rank Reversal Concepts According to Saaty

The basic structure (of ANP), is an influence network of clusters and nodes contained within the clusters and A cluster impacts another cluster when it is linked from it, that is, when at least one node in the source cluster is linked to nodes in the target cluster. Saaty (2008a, b)

There are two basic words here: influence and impact.

According to the dictionary: influence is the capacity to have an effect on the character, development or behaviour of someone or something, and the same source defines impact as to have a strong effect on someone or something.

As an example, the probability of earthquakes may influence without a doubt the selection of a site and obviously may or may not damage the buildings, that is, it has an impact but we need to estimate it.

Saaty uses the concept of *Feedback* to register impact of one element over another or when one influences another, and in addition, he includes this intriguing comment in (Saaty 2008a, p. 2):

Because feedback involves cycles, and cycling is an infinite process when talking about the priorities of the elements in a network and in the alternatives.

Feedback, as a noun, has two different meanings:

(1) Technical: when the output of a process is used to correct the input. For instance, a temperature regulator sends a signal to the heating unit either to shut it down, to start it or to increase or decrease the temperature source (input), based on the room temperature (output).

(2) It is a critical assessment or evaluation on a certain question to improve it, and this is a definition that interests us.

If the question demands an assessment on something, the enquirer is requesting a feedback; he will receive a judgement related to the question, which can be positive or negative.

However, the ANP formulated as a network is not related to assessments but with precedence.

This simple reasoning shows the fallacy of the ANP, which supposedly gives feedbacks or assessments which according to Saaty relate with the future. Look at this sentence, from the same source:

Feedback enables us to factor the future into the present to determine what we have to do to attain a desired future.

Regarding rank reversals, it is interesting that while discussing the reasons of rank reversal with a learned colleague in AHP, he told these authors that in reality rank reversal does not exist in AHP, because it is the result of a wanted feedback. Even assuming that in relating A with B there could be a reaction of B that is transmitted back to A, it would mean that a decision should be done in B based on the input from A, and this is not related to the rank reversal phenomenon that AHP produces, and that Belton and Stewart discovered back in 1983, and that possible was the first and strong blow to AHP credibility.

This book's authors understand that the debate is still open, however, the fact that rank reversal also appears in most MCDM models, seems to refute the hypothesis that it is due to feedback, whatever it may mean. In these authors' opinion, rank reversal in AHP is produced due to the par-wise comparisons procedure between alternatives. Some authors say that if the alternatives are given independent scores, then there is no rank reversal.

2.3   The Decision-Maker and Scenarios

Decision-making involves a mix of quantitative and qualitative aspects, and fundamentally, the opinion, expertise and knowledge of the DM, who has a fundamental role and provides much-needed advice in selecting the best project, since any mathematical MCDM method is only a support tool to help the DM decisions.

Normally, a scenario has quantitative as well as qualitative objectives, and the key for its successfully solving is to have reliable quantitative data coming from recognized sources, as well as trustworthy qualitative data coming from a rational appraisal of a given objective. Unfortunately, AHP fails in both counts, since it advocates working with preferences even when quantitative data is known, and regarding qualitative data, the DM ignores people experience.

There are many voices against this issue, and even a theorem such as the Arrow Impossibility Theorem (Arrow 1951) is against this procedure considering it a dictatorship.

2.4 Conclusion of This Chapter

It has been demonstrated in a practical way that the hierarchical structure used by AHP is not suitable to model the complexities of a scenario, although it appears to be appropriate for trivial problems. The reason for this inadequacy is the lack of flexibility or inability of the hierarchical structure to consider special characteristics of actual scenarios.

Even Saaty, in the abstract of his paper *The Analytic Network Process,* (Saaty 2008a) says: *Many decision problems cannot be structured hierarchically because they involve the interaction and dependence of higher-level elements in a hierarchy on lower-level elements.* There is little doubt that because of that he developed ANP. The latter indeed is an advanced version of AHP, but its only merit in this sense is that it considers all kinds of relationships, albeit at the price of a lengthy and costly process because of the sharp increase in the number of pair-wise comparisons.

The hierarchical structure is very similar to the so-called direct structure or military structure, with a vertical chain of command, and where decisions and orders are taken top-down; no other inputs are allowed and that was the first kind of organization, back in the early twentieth century. This is the way that AHP structure is normally depicted.

Complexities and difficulties in management made this structure obsolete and revealed the necessity to improve it, which resulted in more complex forms such as the functional and matrix structures.

The matrix structure used in linear programming will be explained in Chap. 6.

Complex Scenarios

3

The modelling effort generally defines the richness of the output.
Alessio Ishizaka

Abstract

It is devoted to examine and analyse the concept of complex projects or scenarios and what this term involves.

Since there are no clear definitions about the meaning and scope of this phrase, this chapter proposes ten scenarios, although necessarily briefly, where different features are present and incorporated in their modelling, and including real-world characteristics, such as different relationships between alternatives, and how these can be expressed by criteria, conditions imposed to alternatives, projects underway, no alternatives and so on, that is it gives an idea of what a complex scenario may be and, most especially, how to model it.

Keywords

Complex scenario · AHP · Projects · Criteria · Modelling · Binary matrix · Inclusive · Exclusive

3.1 What Is a Complex Scenario?

Unfortunately, nowadays scenarios are not trivial ones, but they are complicated, complex and very complex; consequently, we need MCDM methods that are able to treat them, and we have the obligation to develop new methods able to do the job. As Velten (2009) states *It is a genuine task for scientists and engineers to deal with complex systems, and to be effective in their work they most notably need specific methods to deal with complexity.*

Let's analyse now the claim that AHP can solve complex problems. This is inexact in two counts:

(a) As analysed above, the hierarchical approach is not adequate for considering many different types of relationships and conditions imposed by a scenario, and

(b) Because the AHP method is not mathematically modelled to consider different features, conditions and restrictions, present even in not too complex scenarios. As a very simple example, the AHP can't solve a contractor selection problem where any two or more candidates are allowed to be bid together, which is a very common feature in large projects.

However, it is necessary to define first which is understood, at least from the MCDM point of view, by a complex project. These authors have not found in the MCDM literature a definition for it, and consequently, propose their own and consider that to illustrate complexity the best way is through many examples, since there are numerous different factors and not always shared between scenarios. Naturally, these examples don't represent the whole spectrum of scenarios characteristics; however, it is believed that they may help the reader to understand the meaning of complex scenarios or projects, and what is more important, to be able to select the MCDM method that best can handle a situation.

Complex projects are those that, in addition to having the typical scheme of objectives, alternatives and criteria, have some of the characteristics described in the following sections, alone or combined. These sections illustrate with examples some of the characteristics of scenarios that must be inputted in the decision matrix. Several of these scenarios are real, and all of them have been solved using the SIMUS method (Munier 2011).

3.2   Examples of Complicated Scenarios

3.2.1   Many Finite or Infinite Alternatives, Perhaps in the Hundreds or in the Thousands

This is common on large projects such as planning river basins (see Cohon et al. 1973, for a project in Río Colorado, Argentina), oil refineries, transportation and so on.

Alternatives may be inclusive, exclusive or dependent, and these three conditions may appear simultaneously in a large project. As an example, examine this scenario:

Statement of the Problem

A City Hall has three projects to upgrade a city avenue affected by flooding, these projects are

(A) Construction of rainstorm sewerage. Involves collection and storage of rainwater in underground tanks, to be later evacuated to a nearby river,

(B) Street lighting. Includes renovation of existing electric ducts for the lighting system for the new LED luminaires, as well as electrical conduits for LED traffic lights,

(C) Paving. Constructing of asphaltic carpet, as well as new sidewalks, and installation of poles for traffic lights.

Data Provided

The three projects *must mandatorily be selected*, and their importance needs to be sequential, that is A > B > C, and this has to be established in the mathematical model, otherwise, the method might select only one of them and even establishing for instance, that C > A, and thus indicating that paving has to precede sewerage and even not selecting project B. In this example, the three alternatives are inclusive, since they can **coexist**, but also there is a *dependency or precedence*.

Discussion

In the real world, we can appreciate often how the no consideration of a precedence like this may show us how our tax money is wasted, when just a months old pavement is broken, to install or build something that should have been performed before.

3.2.2   Many Quantitative and Qualitative Criteria and Other Characteristics

Examples of some special characteristics are as below:

(a) *Dual criteria.*

That is, when the same criterion calls simultaneously for maximization and minimization, which is usual feature in many projects. These criteria may be in linguistic or in cardinal form, or in crisp values derived from fuzzy logic. See the following example pertaining to a dual criterion related to water supply.

Assume a housing development to build three types of dwellings, that is, same types of houses with differences in the interior floor surface and plot size, like backyards, garages and so on.

Statement of the Problem

Suppose the following values (Table 3.1) for water consumption per dwelling, considering *number of people in each one.*

Notice that both criteria have the same label and the same performance values, however, the mathematic symbols and quantities are different. The '$\leq$' symbol

Table 3.1 Water consumption by type of dwelling

	Type 1	Type 2	Type 3	Total litres/day
Water consumption (litres/day-house)	1297	1573	1796	$\leq$ 120,000
	1297	1573	1796	$\geq$ 100,000

means *less or equal than*. The '≥' symbol means *greater or equal than*. The first one is used when the criterion calls for maximization, while the second corresponds to a criterion calling for minimization.

Consequently, the first criterion reveals that the total water consumption per day for all dwellings *must be as a maximum* 120,000 litres. This could be, for instance, due to the size of water trunks as well as for avoiding water waste.

The second criterion is indicating that *as a minimum*, water consumption for all dwellings should be 100,000 litres per day.

This comes from a limit for minimum water consumption per person and per day, established by World Health Organization (WHO). This example shows how resources and their restrictions are considered when modelling.

Discussion

Therefore, the dual criterion is indicating the software that water availability and consumption must be within a certain range, and the model will select an intermediate value. These limits are important because partially they influence the number of total houses to build and how many of each type. That is, an apparent innocent restriction may be key in the whole undertaking and influencing other parameters such as the maximum amount of funds for each type of dwellings.

Another important issue is that criteria may be formed by their performance values – or attributes – expressed in integer, decimals or binary format and/or by mathematical formulas. The last is illustrated continuing with the same example.

The above values per type of dwelling could be expressed using this formula:

$$\text{Total water consumption} = \sum_{1}^{n\,\text{dwellings}} \left(\#\,\text{of persons per dwelling}\right)\left(\textit{Min. water con-}\right)$$

sumption per person) (# of houses of each type), and written in the corresponding cell of the matrix.

(b) *Restrictions may be conditioned to other restrictions in other criteria*

Following the water supply example, the maximum amount of water available *could be a function of another criterion,* such as *Capacity of water wells.* To express this condition, it is possible to link by a mathematical formula the amount of water consumed to the amount of water available in the well. Even the operation could be reversed, by determining how much water is available to each dwelling as a function of the total available from the wells.

This is some sort of sensitivity analysis, to determine how sensible is the supply of water due to seasonal variations of water in the well, but logically, it must be performed before construction and again, it may alter results.

(c) *Criteria may have or not restrictions.*

Example: A subjective criterion such as *Increasing people welfare doesn't have any restriction.* It is obvious that there is neither a low nor a high limit. However, another related quantitative criterion such as *Minimum welfare check* could be established.

3.2.3 Only One Alternative or Option Must Be Selected

In some cases, a production for something complicated takes place in different locations and even in different countries, and then, each place constructing a different component. This is the case of some Airbus aircrafts with plants in the United Kingdom, Germany and Spain, with the assembly plant in France. In this case, it could be necessary, at the beginning of planning, to determine a site for each different part, with several potential sites.

A different problem arises when a client wants to have everything concentrated in one plant, which could be in various sites, and the model should replicate this condition.

Statement of the Problem
A European multinational in the cosmetic business wants to install a plant to manufacture certain products for domestic consumption and for export. Assume that three locations are potentially considered: Bremen, Barcelona and Turin. This not-so-complex scenario has many criteria such as availability of electric energy, price of electric energy, available manpower with experience in these products, nearness to resources, transportation costs to domestic, regional and overseas markets, taxes, labour laws and regulations and so on.

Data Provided
The conditions are *only one site selected*, because the firm is *not looking for a ranking* either, and then the *result must be in binary format*, that is (1) for 'Yes' for the selected site, and (0) for 'No' for the others. Then, it is necessary to instruct the model to consider these demands and produce a binary result.

Discussion
This is a very common problem and the solution gives only one location. However, it can also be that in case where there are several locations, two of them are assigned a '1', then, there is a tie. This circumstance can be solved in different ways. For instance, in the SIMUS method, there are two solutions with different scores for alternatives, but giving both the same ranking. If there is a tie using the first method, most probably the scores of the second solution will indicate the best solution.

It may be that even in this second solution there is a tie. In this case, the magnitude of the shadow prices will indicate the preferred solution.

Another way is to perform a sensitivity analysis. This will reveal the robustness of each alternative, and from there, it can be chosen the more robust.

3.2.4   Considering Timing

Statement of the Problem
In a portfolio, determine what project to execute when not all projects start and finish on the same dates, that is, each project is subject to a schedule. In addition, not all projects have the same durations in years.

Data Provided
In a set of projects with a total duration of 5 years, each project differs from another in the number of years to complete, in the starting and finishing dates and in the rates of execution in percentage in each year. In addition, there is an annual budget that must be respected, and total funds available for this plan are 20% short of the total amount to execute all projects.

Considering all these circumstances, the owner needs to know the projects that must be selected, with the objective of minimizing the investment and considering the available funds.

Approach: The Initial Decision Matrix
The criteria must be formed by *two different matrices*; one is the normal one with its respective criteria (cost, benefit, manpower, equipment, contamination, etc.).

The second *has as many criteria as numbers of years* to develop the complete portfolio, and each criterion corresponds to one specific year, therefore, in a five-year plan, starting in 2020, this is the first criterion, followed by criteria 2021, 2022, 2023 and 2024. Therefore, the *alternatives are subject to these two sets of criteria*, and during each year, there are different performance values for each criterion due to the different construction rate.

The initial conditions are then expressed in the different criteria. Consequently, for each year or criterion, the performance values are the percentages of completion for each project, and it is also specified the budget for that year. This annual budget is the sum of percentage of completion times the budget, both for each project. If we consider 'n' projects, its mathematical expression is:

$$\sum_1^n \text{Percentage of completion} * \text{budget for each project} \qquad (3.1)$$

Discussion
The scenario may still be a little more complicated, if, for whatever reasons, the firm needs only binary results, which requires to add more *binary criteria*, or adding another quantitative criterion regarding the *Net Present Value* (NPV) or *the Internal Rate of Return* (IRR) for each project, which considers the annual investments and the corresponding discount rate per year of construction.

3.2.5   Scenarios That Include Projects Already Underway

This is a very common scenario. In some cases, in addition to new projects, it is necessary to consider that there are already projects underway and at different degrees of completion. Consequently, the model must be instructed to incorporate those old projects; they are subject to the same set of criteria as the new ones, but, in addition, they also *must be in the final ranking,* that is a condition that does not apply to the new projects. In this way, all resources like funds, manpower, transportation and so on are applied to the new projects and also to complete the projects underway.

Statement of the Problem
A national government agency is considering taking advantage of the country's orography, due to the existence of a range of high mountains, for developing a series of hydro-electric projects.

It has identified seven sites with diverse conditions pondering access to the sites, environment, wildlife, population in the area and so on. Therefore, it requests for an MCDM process to determine the best locations in a ranking.

Data Provided
This is a normal problem; however, out of the seven identified sites, there is another site where the construction of a hydro dam is underway and in an advanced stage.

Therefore, it is necessary to complete it and then the need to *incorporate this project together with the rest, under the condition that it must be in the final solution.*

Engineers also found that in two of the seven sites there are conditions for building two dams in series in each site. That is, the water discharged from the first one located at a higher altitude, after generating energy, can be dammed at a lower level and generate additional energy. Consequently, for each of these two locations, there are two projects that are independent, or inclusive, because both can be built; however, if both are selected, the lower one is conditioned to the construction of the higher, although the opposite is not true.

Consequently, the two projects in each site have to be inputted into the model, but also considering the dependency of the second to the first.

Discussion
This arrangement can be seen in the Itaipú Dam in Brazil, and the Yaciretá Dam in Argentina, although they are not dams built in mountainous but in plain terrain; the second, being down water from the first. Both dams are on the Paraná River, and then, the height of the water in the lake behind the Yaciretá Dam depends on the discharge from the Itaipú Dam. Both are independent or inclusive projects, in the sense that both can be built and worked independently, but in this scheme, they are also subject to dependency.

3.2.6   The Alternatives Are Not Projects but Time Periods

Statement of the Project

There is a population cluster formed by a central city, surrounded by many small villages that are situated in an area traversed by many creeks, streams and ravines, and all connected by bridges. These bridges are of several types, different ages and are in various maintenance conditions for business traffic. There are ten bridges, however, number one bridge will be demolished and not replaced.

The City Hall in the central city needs to establish a plan for bridge repairs and care, based on the reports of expert engineers about the mechanical condition of each bridge, as well as complying with annual budgets with funds coming from three different sources, the city, the province and from the national government. This is, in a ten-year plan, and all bridges need to be upgraded in that period. However, for many reasons, as well as for bridges condition, the work has been scheduled in three terms: immediate (Imm.), short term (1–5 years) and long term (6–10 years), along the 10 years.

Data Provided

Each bridge needs to be assigned in which of the three terms it will be upgraded.

The final result must indicate the schedule (Table 3.2):

Unit values indicate the period selected.

Therefore, bridge 2 is scheduled for the long term, bridges 3 and 4 for immediate repair, bridge 4 for the long term and so on, and this solution is optimal since has been solved by Linear Programming.

Discussion

This scenario calls for distributing the total job to be done on the nine bridges along a 10-year period, and then, assigning each one to a certain period, and considering that for each bridge, there could be different assignments depending on the urgency of the repairs and maintenance needed.

3.2.7   Determination of Alternatives as Conditioned by Existing Facilities

Statement of the Problem

Most MCDM projects work with a finite number of known alternatives. However, in some scenarios, these *alternatives are not known* because they are in a very large number, and any combination of them could be used. The proposed project consists in determining a *feasible set of alternatives* as a function of existing urban facilities to build a highway between an airport and a city downtown. See Fig. 3.1.

Data Provided

In this case, no highway is envisaged, but it is necessary to build a rapid road route, with synchronized traffic lights, to allow for a fluid transit, between an airport located in the outskirts of a city and its downtown. There are many different forms

Table 3.2 Final schedule for bridge repairs and maintenance

Bridges	Bridge 1			Bridge 2			Bridge 3			Bridge 4			Bridge 5		
Type of policy	Imm.	1-5	6-10	Imm.	1-5	6-10	Imm.	1-5	6-10	Imm.	1-5	6-10	Imm.	1-5	6-10
Periods selected			1			1	1			1					1

Bridges	Bridge 6			Bridge 7			Bridge 8			Bridge 9			Bridge 10		
Type of policy	Imm.	1-5	6-10	Imm.	1-5	6-10	Imm.	1-5	6-10	Imm.	1-5	6-10	Imm.	1-5	6-10
Periods selected	1				1			1		1			1		

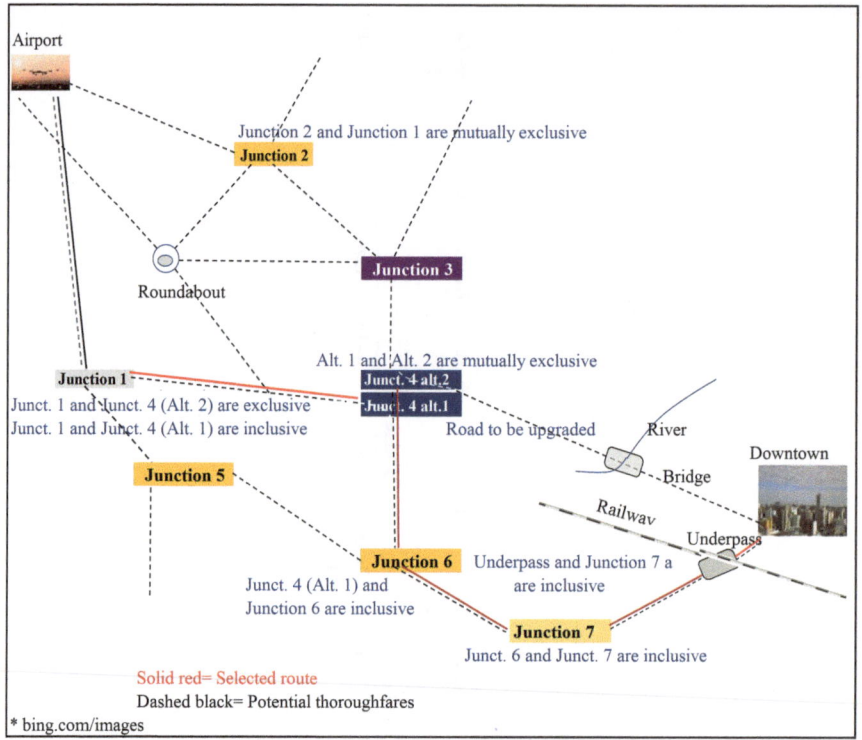

Airport

Junction 2 and Junction 1 are mutually exclusive

Junction 2

Roundabout

Junction 3

Alt. 1 and Alt. 2 are mutually exclusive

Junction 1

Junct. 1 and Junct. 4 (Alt. 2) are exclusive

Junct. 1 and Junct. 4 (Alt. 1) are inclusive

Junct. 4 alt.2

Junct. 4 alt.1

Road to be upgraded   River

Bridge   Downtown

Junction 5

Railway   Underpass

Junction 6   Underpass and Junction 7 a

are inclusive

Junct. 4 (Alt. 1) and

Junction 6 are inclusive

Junction 7

Junct. 6 and Junct. 7 are inclusive

Solid red= Selected route

Dashed black= Potential thoroughfares

* bing.com/images

Fig. 3.1 Determination of the best route subject to existing facilities

to drive from the airport through many avenues, roundabouts or intersections, bridges, boulevards, tunnels, viaducts and so on.

The problem consists in *fabricating a route*, taking advantage of the existing road network. It is necessary to determine the road that can be formed by the set of avenues and round points. There could be *several inclusive and exclusive* avenues and round points, tunnels and bridges. The optimal route will be automatically defined in the result as a series of ways forming a continuous line between the two points.

The problem may be solved using a binary matrix (Table 3.3), establishing for each crucial point or intersection *inclusive and exclusive relationships*, even within a same intersection, where there could be two or more options or alternatives.

It is considered interesting to reproduce this case, which was borrowed, with permission from the book *Strategic Approach in Multicriteria Making – A practical guide for Complex Scenarios* (Munier et al. 2019), Springer, NY.

The matrix shows the alternatives or options materialized by junctions, round points and bridges in columns (facilities), while the criteria are in rows. Observe that in the upper matrix the criteria are restrictions between junctions, that is, inclusive and exclusive. In the inferior matrix the criteria refers to conditions to evaluate alternatives, i.e., the normal procedure.

- Traffic delay,
- Expropriations,

Table 3.3 Initial two interrelated matrices modelling the problem

Road junctions ID	Junction 1 3,569,780	Junction 2 2,256,920	Junction 3 1,096,362	Junction 4 Alt 1 3,045,896	Road upg. and bridge 7,329,980	Junction 4 Alt 2 3,256,980	Junction 5 282,300	Junction 6 2,569,820	Junction 7 459,300	Underpass 2,236,785	LHS		RHS
Minimize investment													
a Junct. 1 and Junct. 4 Alt. 1 Inclusive	1			1							2	=	2
b Junct. 4 Alt. 1 and Junct. 6 Inclusive				1				1			2	=	2
c Alt. 1 and Alt. 2 in Junct. 4 Exclusive				1		1					1	=	1
d Junct. 6 and 7, Underpass. Inclusive								1	1	1	3	=	3
e Junct. 1 and Junct. 2 Exclusive	1	1									1	=	1
f Junct. 6 and Junct 7 Inclusive								1	1		2	=	2
g Alt. 1, Alt. 1, and Junct. 5 Exclusive				1			1				1	=	1

This submatrix is die conditioning or membership matrix

This submatrix is die normal initial decision matrix

	Junction 1	Junction 2	Junction 3	Junction 4 Alt 1	Road upg. and bridge	Junction 4 Alt 2	Junction 5	Junction 6	Junction 7	Underpass	LHS		RHS
Traffic delay	52,611	39,845	42,989	198,996	27,230	88,500	275,698	84,620	55,700	120,000	511,927	≥	MIN
Expropriations		358,930	289,000	450,000	256,000	100,000		1,236,589	456,987		2,143,576	≤	MIN
Signalling	3,590,000	256,420		102,000	75,000	245,896	896,321	562,314	569,872		1,593,186	≤	MIN
Rainstorm sewerage	125,699		796,354				1,236,542				125,699	≤	MIN
Lighting	425,800	228,963	89,745	242,130	250,000	259,862	459,863	236,987		231,842	904,917	≥	MIN
Earthwork		526,930		450,236	345,660	70,200			1,236,987	351,844	1,919,065	≤	MIN
Sum of additional investments	963,111	1,411,089	1,218,088	1,443,364	953,944	1,754,260	2,868,425	2,120,513	2,319,548	351,844	7,198,379	≤	MIN
Total investments	4,532,891	3,668,009	2,314,450	4,489,260	8,193,068	5,021,240	3,150,725	4,690,233	2,778,848	125,896	16,491,331	≤	MIN
	1	0	0	1	0	0	0	1	1	1			

- Signalling,
- Rainstorm sewerage,
- Lighting,
- Earthwork,
- Sum of additional investments,
- Total investments.

Notice that all criteria call for minimization, but using the '≤' operator, except for criterion *Traffic delay* that uses the '≥' operator.

The criteria with the '≤' operator indicate maximum values for *Expropriations, Signalling, Rainstorm sewerage* and so on, and its purpose is to establish a minimization of the maximum amount of funds budgeted for those tasks. The respective funds available are shown at the very right of the matrix, in RHS column.

The '≥' indicates the opposite, that is, for criterion *Traffic delay*; it informs the mathematical model that the delay must be greater than a minimum. This appears to be non-sense, because nobody wants them; however, it is based on practical considerations, since statistics show that always there will be a delay, even when it is small. This is similar to establishing that contamination has to be greater than a minimum, and this is correct, because all anthropogenic activities generate contamination, even with the simple act of breathing.

The upper matrix is a binary one, and it is here where all dependencies are included. For instance, observe that Junction 1 and Alternative 1 in Junction 4 are inclusive and that Alternatives 1 and 2 in Junction 4 are exclusive since both can't be chosen at the same time.

A blue solid row, at the bottom of Table 3.3, shows the result of this problem by a series of '1 s'. Notice that the sequence of these unitary values determines the route.

Discussion

In this example, it is interesting to observe how the two matrices interpret real characteristics of the scenario, concerning relationships or lack of them, and minimum and maximum values. If they were not considered, the result could be completely erroneous. It also illustrates the benefits of using binary matrices and/or vectors which permit a wide latitude for representing strong conditionings.

3.2.8   Determination of Environmental Indicators

Statement of the Problem

Given a set of 100 environmental indicators, select a series of the best according to criteria.

Data Provided

The following conditions must be met:

(a) Determine the size of the final set, maybe 20 or 25 indicators; the size is defined at will by the DM.
(b) The final set must contain as much information as possible from the initial set of indicators.
(c) Each indicator of the final set must comply with a number of criteria established by the DM.
(d) The DM must be able to obtain different final sets for each number of final indicators.
(e) The different solutions must be optimal.

Discussion

This problem requires the use of entropy, computed for each indicator of the initial set, then, calculating a weight for each indicator, and finally using a MCDM method to select the best. The problem may be far more complicated, for instance, it can be required that the final set of indicators incorporate the maximum amount of information contained in the original data. It can also demand that the DM be able to fix at his/her will the size of the final set of indicators. Other requirements could be that each indicator must comply with at least a minimum number of criteria.

3.2.9 Given a Large Amount of Inclusive and Exclusive Projects, Select a Series of Projects That Satisfy Several Specific Requirements

Statement of the Problem

Assign infrastructure projects in a large city and its extensive suburban areas with several projects in each one, as well as projects that must be shared for several municipalities.

Data Provided

(a) Each suburban city has its own projects to be developed,
(b) Projects must be assigned in a fair manner between all cities,
(c) Each alternative must satisfy a number of pre-established criteria,
(d) Some projects, such as sewerage, must serve more than one suburban city,
(e) There is a budget for each city.

Discussion

This scenario is rather complex due to the tough requirements imposed, and because the different alternatives should comply with a series of strong restrictions, involving social, budget and political aspects. To solve, it is necessary to work with binary as well as normal matrices.

3.2.10 Determination of Volume of Different Products to Be Produced in an Oil Refinery and This Production Linked with the Storage Capacity as well as Transportation

State of the Problem
This is a very complex project since it involves working with crude oil and amounts to be produced as minimum and maximum levels of gasoline, fuel oil and other products. The minimum established because economies of scale and plant operation, while the maximum is due to demand and storage capacity.

The problem includes determining the maximum volume of each final product that can be stored as well as the transportation to distribution outlets.

It is necessary to work with related algebraic formulas instead of numbers as performance values.

Of course, economic and environmental aspects must be included as criteria.

In addition, its complexity lies in the fact that four areas are to be included, that is
1. Crude extraction from wells and transport to the refinery, therefore, there is a precedence, and the necessity to comply with established amounts of crude, to feed the refinery,
2. Determination of quantities of final products to be distilled, such as gasoline, diesel oil, gas oil, tar and so on,
3. Transportation of final products in quantities according to market demand,
4. Determination of number and capacity of storage tanks for final products, to absorb variations of production, market and distribution variations.

Data Provided
Just to give an idea, these are some of the criteria to consider:
    Crude
Extraction cost
Transportation cost
Storage costs
    Demand
Maximum and minimum
Average for each product
Loss for delivery delays
    Products
Production cost for each product
Unit price
Balance between production, storage and transportation
    Storage of final products
Costs for each product
    Transportation
Maximum capacity/day

Cost for each product
Products prices
Demand for each final product.

Discussion

In this scenario, there were four operating and independent areas, with the output of one being the input of the next. Consequently, all of them must operate keeping a balance. For instance, production of products must satisfy demand; since this is not constant, and then, there is a necessity of 'buffers' or storage tanks. If production is higher than demand and surpasses storage capacity, it affects the whole manufacturing chain from crude extraction to delivery of final products to the market in the amounts needed for each one. This scenario needs working with complicated algebraic formulas to keep this balance, and those formulas have to be inputted in the decision matrix.

It is not new, since it is common in oil refineries, which, since the 1960s, have been solving it using Linear Programming.

3.3   Conclusion of This Chapter

These ten examples show that there are certain issues that are shared by many projects; they are as follows:

- Large number of alternatives and criteria,
- Not independent criteria,
- Necessity of working with resources,
- Maximum and minimum simultaneous restrictions to resources,
- Relationships between alternatives including dependency, inclusivity, exclusivity and obligation to consider two alternatives,
- Joint ventures,
- Many IF…then…conditions,
- Necessity to consider projects under execution,
- Mandatory selection of a project in the final ranking,
- Performance factors from algebraic formulas,
- Necessity to consider timing of different projects for execution,
- Considering existent correlation between criteria and so on.

Since there is not an accepted definition of complex projects in MCDM, this chapter aimed to demonstrate using examples how a complex problem is, and from there, one can realize that no hierarchical model, including a goal, criteria and sub criteria and alternatives, as much complicated as it might be, can be used to solve them. Consequently, it is believed that an attempt to solve any of these problems through AHP is unrealistic, and then we can safely reject the claims from many practitioners that AHP is able to solve complex scenarios.

Rationality of the AHP Method

4

> *The AHP is the only accurate and rigorous mathematical way known for the measurement on intangibles. It is not going to get old for a long time*
> Thomas Saaty
>
> *Not everything that counts can be counted and not everything that can be counted, counts*
> Albert Einstein

Abstract

This chapter examines the fundamental aspects of the AHP method related with its procedure, especially its preferences and their quantification, its relationship with reality and its appreciation of people's problems originated by projects, that is, the chapter questions its logic, judgement and rationality.

It is explained here the reasons that justify the criticisms that this method has received since its beginnings, decades ago, and that have not subsided.

The many different negative evaluations are mostly based in its lack of a mathematical base, as well as the use of the very debatable pair-wise comparison method, and which is worse, to its ad hoc table to quantify preferences. The AHP method originates many questions from practitioners; some of them are detailed and analysed in a small but realistic sample.

Keywords

Preferences · Pair-wise comparisons · Preference quantification · Mathematical base

N. Munier, E. Hontoria, *Uses and Limitations of the AHP Method*, Management for Professionals, https://doi.org/10.1007/978-3-030-60392-2_4

4.1   Introduction

There is a typical question from practitioners: How can stakeholders trust in the preferences of a DM or group of DMs to select a project, when another DM or a different group may get different solutions or rankings? Which of the two solutions should they accept and on what grounds? Which is the yardstick to measure the goodness of each solution?

These questions don't have an answer.

Naturally, we can accept that all DMs and groups are knowledgeable people, completely unbiased and coherent, but it does not preclude that they may have different preferences, know-how and experience that they apply when confronted with the same issue. The problem is that preferences are subjective, and then, with neither mathematical support nor reasoning, consequently, why the stakeholders should accept the preferences of a determined DM? Probably, it can be said that preferences are meaningful for the DM, but the fact is that stakeholders are interested on a project and not on the DM feelings and moods.

Let's assume that a medical doctor, a dentist and an engineer must justify why they took certain actions in their respective areas. The medical doctor and the dentist can say, and show, that they were based on the patient history, on the x-rays and on diverse analysis. The engineer can show his computations, bibliography and drawings, that is, the three of them have solid proofs that sustain their decisions. Now, if we ask a DM why he decided that quality is three times more important than price, which will be his argument? That he followed his intuitions? What the stakeholder's reaction would be when receiving this answer?

This simple comparison illustrates the absurdity of the pair-wise comparison and the lack of scientific basis for AHP.

4.2   The Nature and Reasons for Criticism

The word *criticism* may have different definitions regarding the object that receives the critique.

In this book we criticize a MCDM method, the AHP, trying to point out its weak features from a concrete point of view, and from its structure, however, we don't venture into its mathematics.

The latter is an AHP area that has been active for more than 30 years, practically since its publication, back in the 1980s, and was further addressed in hundreds of technical papers by mathematicians, MCDM researchers and professors, and in both ways, that is, defending and criticizing the method; it is believed that everything that was meant to be said about it was already expressed, and then, we consider that it is not necessary to add more demonstrations to an already abundant literature.

In this book, the critique is oriented towards the lack of common sense in AHP and its inability to model and solve real-world problems. It appears that in this method practical and tangible aspects have been overlooked, by prioritizing the psychology of the DMs and their intuitions, over reliable and available information, something important indeed, but not in the extent attributable to the DM.

These authors firmly believe that the DMs are the most important component in any MCDM process, but they also think that the DMs expertise, know-how and good sense, should be put to work at the end of the process, in a bottom-up approach, when they can express their preferences, with the support of a result objectively and mathematically obtained.

4.3   Briefing the Objective of This Book

As mentioned, other than the criticisms from Belton and Stewart and Dyer, there are many more aspects than have been addressed along the time, with no sound answers either from Saaty or/and from AHP supporters, or practitioners.

For this reason, these authors thought that it would be convenient to organize, comment, discuss and classify those issues that were and are subject of criticism, and this is the genesis and purpose of this book, something that the authors believe was never done before.

That is, its objective is not to belittle or degrade the method, but the authors consider that their obligation as professionals is to ethically denounce aspects that are wrong, hoping that they can be fixed.

4.4   Why This Criticism?

The AHP is the MCDM method with the widest usage and diffusion in different fields and enjoys approval and popularity from many practitioners, but has also been object of negative and harsh reviews from reputable scientists, since its inception and in two different aspects: the soundness of its mathematics and the procedure, especially the preferences and the evaluation of those preferences.

His creator, Thomas Saaty, and AHP experts Luis Vargas and Patrick *Harker* have tried to dispel this criticism, arguing that *this criticism arises out of a lack of understanding of the theory underlying the AHP (Harker and Vargas 1990)*, or that it was not applied properly, but apparently, they didn't succeed, because the debate remains well alive till our days (2021), as well as new critiques to the method that surface in papers and books.

Decades ago, Belton and Stewart (2010) discovered the controversial rank reversal phenomenon in AHP, and in addition, Dyer (1990a, b) criticized the method, which provoked the Saaty (1990) reaction and answered with the above mentioning sentence. These comments offended Dyer, who responded to the rebuttal. Those negative comments from Belton and Dyer and Saaty's defence remain, after 40 years, unsolved, in spite of several researchers expressing their opinions in one way or the other.

When, as in this case, criticisms and the interchange of hot communications – made public – come from highly respected professionals, the mentioned no-scientific explanation about colleagues not understanding the mathematics and the structure of the method obviously does not qualify as a professional and credible answer; quite the opposite, it shows that there are no rational explanations.

4.5   Which Is the Base for AHP?

AHP is based on *pair*-wise *comparisons* between criteria and also among alternatives. This aspect will be further discussed in the corresponding section. Pair-wise comparison, or the *Law of Comparative Judgement,* was a method developed by Louis Thurstone (1927) and is considered a very good tool for analysing a problem, but it is also pondered as not adequate for decision-making.

AHP is mainly based on this tool that was designed to compare subjective factors, and Saaty added his Fundamental Scale to measure how better or more important was something related to another. However, how to measure the importance of an emotion or an intuition over another? This remains unexplained or pretending that following certain rules this importance can be appraised.

If this procedure is applied to projects with hundreds of quantitative and qualitative criteria, one can understand the irrationality of this approach, even when trying to compare, for instance, not so subjective elements such as how much important is safety than manpower.

If criteria affect a large number of people, for instance, when it is necessary to evaluate the relocation of thousands or even millions of people, as happened in the Three Gorges Dam in China, how can a DM put a value of importance to people suffering and problems because of the relocation, against another criterion such as the benefit of building a new road?

For these and many other reasons, the authors believe that the preferences system has no room in MCDM problems, although maybe it is fine for some trivial situations, where the consequences of the decision fall on the person who decides. It is not casual that articles advertising or explaining the AHP choose trivial problems such as purchasing a car, selecting a restaurant for dinner or an apartment for rent, but not to real-world projects, which affect many people or the environment, or compromising hundreds of millions of dollars.

If it is true that decision-making involves a good deal of subjectivity, it is believed that it is not the field for applying psychological theories and assumptions as used in AHP, to justify, for instance, the maximum number of criteria, sub-criteria or the *Fundamental scale.* It is preposterous and unreasonable to assume that in a problem such as the industrial location, for instance, the best site may be decided by preferences.

Preferences must be taken at the very end, that is, post-result, where the DM can express his preferences on the result achieved by the mathematical model, which only purpose is to support the DM decisions. These must be based on his experience, knowledge and analysis of the situations they may arise if those results are applied. This is one of his most important functions.

Based on their know-how and expertise, the DMs may modify data, add or delete alternatives, criteria and performance values or reject the result, if at their judgement, even if the result is mathematically sound, it is not convenient because the model did not take into account certain subjective factors, that were either omitted or not considered, for whatever reasons.

DMs must forecast the behaviour of the best project in the future, when subject to unknown but foreseeable upcoming conditions. These are the kinds of preferences that they can apply to a problem, not to establish preferences by intuitions at the very beginning and without any data that can support them.

They can also select a second-best alternative due to the sensitivity of the best solution to a certain criterion change, by checking the robustness of the solution.

As seen, it appears to be more reasonable to establish preferences on known results, obtained by a mathematical procedure, than to start the analysis with an initial matrix with values that have been altered or worse, estimated, guided by intuition. The first is a bottom-up approach instead of the second, which is a top-down approach, and used by AHP and other methods.

Another very serious drawback of the AHP method – albeit it is fair to mention that it is common to most MCDM methods – is the procedure to perform a sensitivity analysis. In this very important and unavoidable step, the AHP analysis is based on the dubious validity of debatable data, as are the trade-off values elicited by personal preferences, and their variation to determine how the output or ranking performs, because the trade-offs are not fit to be used to evaluate alternatives.

This aspect is treated in the corresponding section. However, this is only a part of the problem, since AHP relies on the variation of only one criterion, and in addition, arbitrarily chosen, when in reality should be considered the simultaneous variation of all intervening criteria.

4.6   Small Survey Involving MCDM Practitioners

One of the authors (Munier) has been a member during years of a scientific forum called *ResearchGate* (RG), where continuously thousands of practitioners, researchers and professors of several disciplines interchange ideas, formulate technical questions and get answers on aspects of their interest; those with know-how on a particular matter normally share their knowledge and experience in order to help, by clarifying doubts, giving advice and recommending publications, and sometimes helping a practitioner on a certain procedure or method.

This is a strictly scientific forum and freely accessible to be read and participate by anybody. It is believed that it helps many people when receiving, without any pre-condition or interest, the advice, the expertise and the transfer of knowledge from many people around the world addressing many different kinds of problems. It acts as a large classroom with many students and also many professors, responding and clarifying a myriad of different questions and suggesting procedures for different MCDM problems.

Munier thought that this forum could be used as a laboratory, some sort of a multi-dimensional and massive survey, which logs during years of questions and opinions from practitioners, users and researchers, materialized in thousands of different enquires, and including recursive doubts on AHP, as well as on other MCDM methods.

As can be understood, it is a rich depository of thousands of questions as well as of answers, advices, comments and discussions on different technical issues. Some questions are responded by different participants, and sometimes they are even contradictory, which give the asker a big latitude for him/her to learn and take a sound decision.

In the MCDM area, there are questions regarding several of its methods, such as PROMETHEE, (Brans and Vincke 1985), ELECTRE (Roy 1968), TOPSIS (Hwang and Yoon 1981), VIKOR (Opricovic and Tzeng 2004) and others, as well as the application of fuzzy logic, Zadeh (1965) but by far its largest percentage is on AHP and ANP. Most probably, this largest percentage of communications obeys to the large diffusion of AHP, or perhaps, as the analysis of the questions suggests, because, by a large amount, the AHP originates more doubts than the other methods.

Along the years, Munier repeatedly posted in the forum the same issues on AHP that are now in this book, and asking for responses, denials, rebuttals or technical explanations, and even requesting to correct him. Unfortunately, he never received answers, never received rebuttals, never received arguments defending the method, which appears to indicate that practitioners and colleagues don't have answers for his questions or refutations for his comments, and hinting that their silences indicate approval.

It seems that people recognize the shortcomings of this method, but do nothing or even discuss them, or simply are not interested, and it is really difficult to imagine the reason. These authors only hope that this book may generate a healthy acknowledgement from AHP users, which can help them to know how to handle or not, their problems when using AHP, or if they have to look for more convincing and realistic methods.

It is interesting to make some counting and thinking about the most frequent questions regarding AHP. This is a sample of verbatim questions formulated by practitioners.

Most questions below are about how to:
1. How to consolidate results from experts or people regarding preferences?
2. How to deal with quantitative values?
3. How to deal with especial characteristics of the scenario?
4. How to validate results?
5. How to select criteria considering that it is recommended a maximum of 9?
6. What happens if the transitive product is larger than 9?
7. How can I get rid of the comparison scale?
8. Which result is better from AHP/ANP comparison?
9. How to deal with the relationship between two mutually exclusive independent variables on dependent variable in one model?
10. Is it correct to consider that humans are rational in decision-making?

11. Which is the appropriate model for pipeline geo-hazard assessment? (Except AHP, ANP, EIA)?
12. Why multi criteria decision-making techniques, like AHP, are not common for allocation of healthcare facilities?
13. Can you please explain how you combine real data with experts' opinions using AHP?
14. Could somebody explain why AHP is used, other than its 'easiness' to use?
15. Is it rational in compensatory models that when modifying a single criterion weight, all the other weights modify proportionally?
16. How do I convert quantitative data (result from a questionnaire) into an AHP hierarchy structure?

Of course, this is only a small sample and not serious conclusions can be extracted from it, but even from its size, it can be realized the variety of questions, doubts and concerns from users.

If a method that has been around since the 1980s and used in thousands of examples originates these questions, then it is evident that people find that it does not satisfy their requirements.

Making a few comments on each of these questions:

Number 1: Ask for a procedure that apparently is not well explained. This question has been formulated several times and by different persons. Therefore, it appears that the method produces more problems than solutions. Analysing the answers, it seems that some differ from others, then, the new question is: Which answer should I use? For a method that boasts that it is grounded on sound mathematical principles, this is indeed very strange and confusing.

Number 2: Ask for a procedure that apparently is not well explained. This question has been formulated several times and by different persons. No help was received other than encouraging the user to ponder the problem and proceed.

Number 3: This question confirms one of the most important AHP shortcomings, since it seems that the person asking this query wonders how to input into the model the characteristic he is interested in or detected in the scenario. This is also the main criticism formulated here for AHP and very clearly shows the method's inability to model reality.

Number 4: It is impossible to validate results in AHP or in any other MCDM method, except when the result is physically known, as in the naïve and misleading Saaty example of geometric shapes, which does not represent real situations, because reality is never known.

Number 5: In this case, the practitioners found that they need additional criteria, but the same Saaty advises to use no more than 9, based on psychological findings. Too bad that nobody explains how to solve this. How to proceed then? As seen on diverse occasions, the solution is to limit the number of criteria and sub-criteria. In so doing, it could mean that the alternatives could be not properly evaluated for lack of evaluators, that is, missing to be subject to certain criteria as they should. However, this appears to be a matter not enough important to be considered.

Number 6: Obviously, the practitioner was aware of the absurdity of that scale. Saaty says that there is no limit in this scale, however, he does not explain how to proceed.

Number 7: The user is asking how to get rid of the main component in AHP. This could mean that he is fed up with some many comparisons or that he sees no logic in it.

Number 8: Apparently, the user does not know which result is better. It appears that nobody told him that results will coincide if criteria are considered independent in ANP. It also could mean that the user does not realize that normally results are different.

Number 9: This question has no answer because, in AHP, those conditions can't be modelled. This illustrates the failure of the method to model according to the scenario characteristics.

Number 10: The user is questioning the very essence of AHP, which assumes that the DM is rational. It would be interesting to learn on what grounds that assumption is based because Saaty does not explain it. The dictionary defines a rational person as *A rational person would look at the facts behind every issue before making a decision no matter how comfortable or uncomfortable it might make him or her feel.*

The definition makes sense, however, the DM in AHP expresses his preferences by intuition, not by considering which is the best. We have to remember that *intuition* is how AHP defines the preferences.

The dictionary defines intuition as *The ability to understand something immediately, without the need for conscious reasoning.*

Therefore, if the DM acts by intuition, he is not rational because he is not reasoning; he is not thinking critically.

Number 11: The user asks for any method, except AHP, ANP and EIA. It would be interesting to know why the user is not taking into account any of these three methods, although EIA is not an MCDM method.

Number 12: The answer is not known, however, if the user asked this question is because he/she noticed this fact.

Number 13: Again, a practitioner is asking how to proceed. No one answered.

Number 14: This author formulated this question; RG metrics indicate that it had 225 reads but not answers.

Number 15: This author formulated this question; RG metrics indicate that it had 76 reads but no answers.

It is believed that it is interesting to think about these questions and the notorious absence of answers. Anybody can check these questions by accessing the RG website.

4.7   Conclusion of This Chapter

This chapter explains the reasons by which this book criticizes the AHP method. There are several aspects that can explain this criticism:

(1) One start wondering how a method can be thought as giving reliable results when these are personal and depend on the person doing the analysis. It is like being at the same time judge and part, because the DMs express a preference and themselves judge if it is correct, until a formula tells them that it is not so....... Since preferences are subjective, then, there is no guarantee that another person will think in the same way.

(2) According to many, AHP is grounded on solid mathematical principles. Really, one can admit it when referring to using the eigenvector method, or simulation, or the geometric mean, but it is hard to believe that said assertion may include pair-wise comparison and quantifying the preferences by using a very debatable scale, both without any mathematical background.

(3) There are two aspects that are much heralded in AHP literature:
   1. The DM is assumed to be rational (Triantaphyllou 2000), that is, a person that reasons, that analyses, and therefore, acts judiciously and wisely,
   2. The DM acts intuitively.

Then, there is a contradiction here, since if a rational person is somebody that reasons, but, if this person is guided by emotions, feelings and intuitions, how can he be rational? Curiously, Rozann Saaty (2016), in page 3, declares that AHP is free of paradoxes.

The chapter concludes with a small survey about AHP done using a scientific forum. There are many comments questioning different aspects of AHP, however, none defending it.

Shortcomings of the AHP Method

5

> *Once you have accepted a theory and used it as a tool in your thinking, it is extraordinarily difficult to notice its flaws*
> Daniel Kahneman

Abstract

This chapter investigates and examines a general criticism of the AHP method and 30 specific subjects, shortcomings and drawbacks. It establishes the procedure as follows:

These subjects have been identified and analysed in depth by these authors, according to their own research and methodology, not using mathematics, but reasoning, common sense and examples. After each examination, opinions, comments and views of different researchers on each subject are transcribed verbatim. Naturally, there are subjects that merit a good deal of attention from many scholars while others have been considered by only one or two, or even by nobody. The latter, in an amount of three, correspond to subjects that have been not addressed in the literature. That is, out of 30 subjects proposed and analysed in this book, 27 have also received attention from other researchers; the remaining three are identified by these authors.

The chapter ends with a table vis-à-vis that expresses in one column these authors' viewpoint on each subject (in general, only a broad opinion), and in another column, expressions from other researchers. Not assessments are made considering both opinions, because the aim of the table is to facilitate the reader to make comparisons, and thus, allowing he/she to extract his/her own conclusions.

© The Author(s), under exclusive license to Springer Nature Switzerland AG 2021
N. Munier, E. Hontoria, *Uses and Limitations of the AHP Method*, Management for Professionals, https://doi.org/10.1007/978-3-030-60392-2_5

Keywords

AHP · Pair-wise comparisons · Weight · Preferences · Modelling · Complex problems · Quantification · Sensitivity analysis · Transitivity · System · Rank reversal · Portfolio · Limiters · Hierarchy-structuring

5.1   General Criticism

This chapter is divided into 30 subjects or sections, each one addressing a shortcoming of AHP. The authors make an analysis of each one, and for clarification, illustrate it with real-world examples. No references or opinions from other authors are included here, as it is normal in technical texts. Instead, there is a second part for each subject with verbatim comments from diverse authors related to the issues discussed. These authors do not participate in these comments; they only post them.

Thus, in each subject, there are two sets of independent opinions, those from the authors and the others from different specialists. These are the opinions, comments and advices of more than 100 renowned scientists, often several with more than one opinion on the same subject.

They may or may not agree with the authors' opinions, and their only purpose is to show the reader how they ruminate on the issues addressed in the first part.

In some few cases, there is no correspondence between the authors' comments and those from experts, because the first are theirs, with their own perceptions, ideas and findings, which it is believed have not been addressed before by these experts.

The authors base their analysis from reading hundreds of papers written on AHP and applications, in having worked with AHP and ANP, and by talking and writing to authors, who normally don't like criticism, except the late Dr Saaty who very kindly answered our questions and concerns on his method, even when they were against it. Unfortunately, his passing away interrupted this healthy interchange of ideas; we hoped to work jointly with Dr Saaty but regrettably, destiny decided differently.

As said, these authors express their points of view on several aspects, and most of them exemplified. However, those are only their opinions, and then, it is considered that it would be useful to contrast vis-à-vis their judgements with those of researchers mentioned in each subject, and then, comparing their opinion to those of many.

No conclusions are posted for each subject and notice that researchers' opinions may be in favour or against of what these authors state.

Observe that there is only one subject on the authors' side (first column), with the general title of each subject, while verbatim words, sentences, opinions and comments are posted on the researchers' side (second column).

The format of this table is replicated at the end of each subject, except, naturally, in those subjects with only these authors' opinions.

Thinking, reasoning, good judgement and reasonableness are everybody's attributes and can be performed by examining the characteristics of the theory and the procedure behind AHP and extracting conclusions, something that is also available to everybody. These authors are just two of them.

The bibliography at the end of the book details all the authors as well as the books or journals where they published their ideas and applications.

5.1.1 Experts Comments and Opinions on General Criticism

Dyer (1990a, b): AHP is fundamentally unsound and flawed.

Harker and Vargas (1990): Much of the criticisms of the AHP are based on a misunderstanding of the theoretical foundations of the AHP.

Ishizaka and Labib (2009): AHP is useful when the DM is unable to construct a utility function.

Smith and Winterfeldt (2004):

(a) While many in the decision analysis community (ourselves included) follow Dyer in believing the AHP to be fundamentally unsound, others (including Saaty, Harker and Vargas) disagree, and the AHP is still widely used in practice today.

(b) While Saaty provided an axiomatic foundation for AHP, these axioms conflict with the axioms of expected utility theory.

Rodrigues **et al.** (2017): The AHP/ANP underlying axiomatic (Smith and von Winterfeldt 2004) is considered by several decision analysts as unsound, being a controversial method in the field of MCDA.

Bana e Costa and Vansnick (2008):

We have addressed the foundations of AHP, by analysing the eigenvalue method (EM) used to derive a priority vector. Our main conclusion is that, the EM is very elegant from a mathematical point of view, the priority vector derived from it can violate a condition of order preservations that in our opinion is fundamental in decision aiding – an activity in which it is essential to respect values and judgements.

In light of that, and independently of the other criticisms presented in the literature, we consider that the EM has a serious fundamental weakness that makes the use of AHP as a decision support tool very problematic.

Asadabadi **et al.** (2018):

(a) Two traditional Multi Criteria Decision Making (MCDM) methods developed by Saaty (Saaty 1986, 1996). However, both methods carry inherent deficiencies that affect the rankings in their real-world applications (Belton and Stewart 2002; Dyer 1990a, b; Ho 2008; Önkal et al. 2009).

(b) The Analytic Hierarchy Process (AHP) and the Analytic Network Process (ANP) are considering the fact that the general form of MCDM does not require many pair-wise comparisons, and it does not concern the decision-maker with computing consistency levels; the question that arises is: Where shall we use AHP and how can we be assured that after all the extra efforts that we put in, AHP leads us to more reliable results than a simple MCDM method?

Kahneman. Cited in Triantaphyllou (2019):

(a) The errors of a theory are rarely found in what it asserts explicitly; they hide in what it ignores or tacitly assumes.

(b) Once you have accepted a theory and used it as a tool in your thinking, it is extraordinarily difficult to notice its flaws.

Salvia **et al.** (2019):

On the other hand, some disadvantages of this method (AHP) include the high computational requirement even for small problems, having a subjective nature and relying on emotions to be transferred to numerical judgements and the increased time and effort demanded bigger amount of pair comparisons (Karthikeyan et al. 2016; Oguzitimur 2011).

Jimenez **et al.** (2004): (Translated from Spanish). One of the weakest points of AHP (Saaty 1980; Jimenez et al. 2004) is the difficulty in working with large problems, that is, with problems with many hierarchy levels (above 4).

Perez **et al.** (2006): Since in almost all applications of AHP, the set of criteria is not fixed ex-ante but is variable and is constructed in accordance with reasons of relevance and simplicity, almost all applications of AHP are potentially flawed.

Zardari **et al.** (2015): Criteria may change their value in space and in time, and then pair-comparison is the worst method for weighting.

Saaty (2008a, b): To make a decision, we need to know the problem, the need and purpose of the decision, the criteria of the decision, their sub-criteria, stakeholders and groups affected and the alternative actions to take.

Comparison

These authors' opinion on this subject	Arguments, observations, judgements, remarks and opinions from MCDM experts, researchers and authors on this subject
AHP has many drawbacks	Eleven researchers articulated their opinions about this subject, using expressions such as:
	Flawed,
	Unsound,
	Controversial,
	Having fundamental weakness,
	Having inherent deficiencies,
	High computational requirements,
	Many of its applications are flawed,
	The method uses the worst method for weighting.
	As a defence, other researcher expressed:
	criticisms originate because of misunderstanding.

5.2 Comparisons Among Criteria

5.2.1 The Pair-Wise Method and Its Application in AHP

AHP uses pair-wise comparisons, a conception developed by Louis Thurstone in 1927. They are good for analysing a project and examining the advantages and disadvantages, when based on a realistic analysis.

A DM can say, for instance, that from the point of view of the company global objective of maximum profit in selecting a project, the IRR is more important than the pay-back period of money invested, and this is a very legitimate preference, because it is probably based on different financial considerations, and related to other aspects of the business. However, it is very difficult to place a quantitative value for this preference, because many things are not measurable, and even if they are, they may apply to different things.

As an example, assume that a DM concludes that the criterion *IRR financial indicator* is more important than criterion *Pay-back period financial indicator*, and he reached this conclusion after talking with accountants, financial officers and company directors. Some of them say that IRR is more important, because it indicates how profitable an investment could be and thus allows for boosting the confidence.

Others say that the *Pay-back period* is more important, since it may reflect positively in the stock market, because the sooner the investment is recovered the better, which allows for more available funds for other investments, which derives in risk reduction, and this, in turn.... and so on.

These are legitimate assessments coming from knowledgeable people who are rational, and then, the DM may have reasons to prefer one over the other. But the decision is not straightforward or black and white, because usually there are strings of direct and indirect consequences that not even experts can be sure about,

Even if the DM has a Master's in Business Administration, how can he make decisions when there are diverging opinions from experts in each area?

Even if he can, and says: *From my point of view the IRR is slightly better than the Pay-back period.* This is a very understandable preference, but what is not reasonable is that automatically, this slight preference receives a value of 3 according to *Saaty Fundamental Scale.*

If the DM analyses now the IRR with the Net Present Value (NPV) and finds that the latter is slightly better than the former, then does it imply that the NPV is nine times better than the Pay-back period?

This is neither credible nor scientific, because there are not elements to justify it; this reveals the lack of common sense in this procedure.

A MCDM problem is not a series of simple preferences subject to the will and intuition of the DM, but a tough undertaking because of the uncertainties and the characteristics of the scenario. For instance, in AHP:

(a) It depends on who is doing the comparison, and when there is a group of decision-makers, it is normal that their opinions diverge. They receive information and apply their knowledge and finally may reach a consensus about which of the two criteria is more important. However, once a decision is taken, how can they reach an agreement about the number of times that one is more important than the other?

How can they quantitatively evaluate the factors that make one of them more important than the other?

(b) It depends on the nature of the comparison. It is obvious that it is not the same to select a car among several makes that select a location for a new industry in three different countries. Why?

Because, besides the different complexity, one thing is to take a decision that only affects the DM; i.e. the selection turns out to be a bad car, the disappointment falls on who took the decision, and it is short-lived; it is static, since when it is over there are no consequences, and sometimes they can be reversed.

In the second case, a wrong decision can cause very large losses and hurt many people, and usually is highly dynamic, because its effect can last for years, as well as having a strong multiplier effect, and are mostly irreversible.

For instance, the decision to relocate people because the land they occupy is needed for a hydroelectric project may not only hurt people but also may have serious consequences that are not even hinted. As a real example, when the Itaipú dam was built in Brazil, many people were offered economic compensation. Trouble was that it was not enough for them to buy land and many had to emigrate to nearby Paraguay, across the river. The migration of so many people had economic consequences to the area, which saw its regional GDP to fall abruptly.

As can be seen, decisions in real world have nothing to do with the psychological capacity of DMs, as Saaty said his method was based on, but on real and tangible effects and also very often, affecting people feelings and may be altering their way of life. The latter is not a bold assumption, because all projects, whatever their nature, relate to people.

(c) Projects involving population. In these projects, there is always the possibility of asking people, through polls and surveys, about certain issues, naturally not technical. However, when people are asked to answer complicated forms and with difficult questions and make comparisons, there is a tendency to give any answer, just to get rid of the pollster.

Does it make sense to ask a dweller how much important (in numbers) is to have a new house built in another place in lieu of the old one where he/she has lived during decades? What about the feelings, the memories, the recollections of people that must evaluate them when comparing with a new more comfortable house? The questionnaire is asking them to quantitatively compare a very valued and dear intangible, with the prospect of a tangible and unknown alternative.

It would be better to ask the people their opinion about how the different projects may affect their life. For sure, people will have a clear understanding of it and will be possibly eager to answer.

In addition, the result of the survey normally goes to the DM who may alter them, according to his preferences or interpretations.

Sometimes the elemental step is omitted, and it is the DM who decides in lieu of the people. As a known expert told to one of the authors years ago: *People need to be taught.*

To be taught what?

Does it mean that the DM (an outsider) is qualified to teach people affected by a new project, on how to face their own problems provoked by it? People know better than the DM about their needs and how their style of life would be changed. The DM cannot possibly know or appreciate how a future highway cutting in two a city can affect their lives, changing their purchases routine, making more difficult for children to go to school, changing for many people a nice view for the sight of a noise barrier and so on.

The assumption of being able to represent the wishes and needs of affected people, or to correctly understand certain risks, or to evaluate aspects that must be addressed by specialists, make that AHP not suitable for large projects, and only partially acceptable in trivial scenarios, where the largest complication is usually the size of the hierarchical structure.

Large projects with hundreds of criteria and many alternatives are not the place to use preferences based on intuitions.

(d) Another incongruity: the fact that criteria relative evaluation may also depend on the professional background of the DM and on his personal and technical knowledge and ideas on certain disciplines. Say, for instance, that the DM must compare criterion *Production* to an environmental criterion such as *Contamination*.

If he is an engineer, then, obviously he will have a good knowledge and experience related to cost, resources, efficiency, manpower, performances and so on, and may be some loose ideas about environmental issues. Therefore, he must compare two different criteria where he has only a very good knowledge of one of them, and ignorant about the other.

How can a comparison be made, if say he is unaware of existing international regulations about noxious discharges into the air, water and soil, and all of them different?

It could very well be, that he does not know that a selected project possibly will not be approved because it can cause international conflicts because the discharge of contaminated waste into an international river. This case is real, and it happened on the Uruguay River between Argentina and Uruguay, due to the installation in Uruguay of a paper mill in the shores of the shared Uruguay River. In this case, it could be that contamination is more important than cost.

If the DM is an environmentalist, then for sure he/she will know about this, and very little or nothing about costs.

(e) There is another hurdle with pair-wise comparisons, because normally, the two compared aspects do not have each one a single scope; each criterion may have different scopes. For instance, assume that we are selecting countries to produce oranges for export, and that two criteria among others are *Country factors* and *Country conditions for export*.

The immediate question is to what aspect the DM refers in each criterion. For instance, *Factors* may be positive and negative and related to adequacy to grow the fruit, and here he needs to consider several different concepts, such as sunlight, soil, rain, quality of fruit, pests, government regulations and so on.

Conditions fort export may refer to several items, such as country location, harbour facilities, export taxes, distance to markets and so on.

Therefore, the DM has to first define what kind of factors he is referring to, as well as what specific facility for export. Then, according to AHP, he can ask: *From the point of view of benefits, which is more important country factors or conditions for exporting?* and has to identify exactly which are the constituents of this pair.

It does not have too much sense to bluntly assert, for instance, that factors are more important than export conditions. A country may produce a delicious, abundant and cheap fruit, but, if it is landlocked and at 2000 km of the nearest harbour in another country, factors importance is low. Conversely, a country may have a lot of opportunities to export because its location in the world, its deep-water harbour and a weak currency, but if it for whatever reasons produces only small quantities of fruit, it is obvious that its factors are more significant than export.

The pair-wise comparison is even more complex, and in addition irrelevant, because not all factors act in the same manner regarding exports. Consider, for instance, that the same landlocked country may produce very large amount of excellent fruit, which is harvested twice a year. In this case, because of the high international value of the fruit and high demand, and superb quality, the company decides that benefits are by far, more significant than transporting it 2000 km., that is, opposite as before.

(f) Regarding consistency of preferences, there is also something that needs to be examined in detail. It is related to determining the extent to which changes in the decision matrix to get CR < 0.10 may affect the weights of criteria. This is some sort of sensitivity analysis where the inputs are the preferences and the output are the weights. For this, we must use some simple algebra.

If the values of the preferences between criteria are known, and if there is perfect consistency, then these same values must hold for ratios between criteria weights, as expressed in formula (5.1).

$$a_{ij} = \frac{wi}{wj} \tag{5.1}$$

where a_{ij} is the criteria preferences and w_i and w_j are the respective weights. To find these weights, we can use the geometric mean.

Table 5.1 shows a matrix built by the DMs based on their preferences:

The equation will be satisfied only in the case of a consistent matrix (see Glossary), as happens in this example:

To compute the weight, the geometric mean formula (5.2) is used.

Table 5.1 Initial matrix from DM preferences

	C1	C2	C3
C1	1	3	6
C2	1/3	1	2
C3	1/6	½	1

$$\sqrt[N]{\Pi\,aij} \qquad\qquad (5.2)$$

Therefore:

$$w_1 = \sqrt[3]{1*3*6} = 2.62$$

$$w_2 = \sqrt[3]{\frac{1}{3}*1*2} = 0.87$$

$$w_3 = \sqrt[3]{\frac{1}{6}*\frac{1}{2}*1} = 0.43$$

Now, we find the ratios between criteria weights.
$w_1/w_2 = 2.62/0.87 = 3$ against a preference of 3 between C1 and C2
$w_1/w_3 = 2.62/0.43 = 6$ against a preference of 6 between C1 and C3
$w_2/w_3 = 0.87/0.43 = 2$ against a preference of 2 between C2 and C3
Observe that the ratios between weights coincide with the preferences of criteria and satisfying Eq. (5.1). That is, there is no error.

5.2.1.1 Sensitivity of Rankings Due to Changes in Preferences

This is a subject that, as far as these authors' knowledge, was not previously studied.
Assume that the DMs have built the matrix shown in Table 5.2 and they find it inconsistent, consequently, following AHP procedure, they want to render it consistent considering CR < 0.1. We want to examine the influence of such a change in the criteria weights.
The preferences are then as below:
C1 = 2 C2
C1 = 3 C3
C3 = 2 C2
Using the Simple Priority Calculator (Goepel 2019) for determining the eigenvector method as in AHP, the DM gets the eigenvalue as $\lambda = 3.14$, and for n = 3 and

Table 5.2 A matrix from DM preferences

	C1	C2	C3
C1	1	2	3
C2	0.5	1	0.5
C3	0.33	2	1

using the Saaty's Table for RI, finds that the Random Index (RI) = 0.58. Then the Consistency Ratio (CR) is determined by using the formula (5.3):

$$CR = \frac{CI}{RI} = \frac{(\lambda - n)/(n-1)}{0.58} \tag{5.3}$$

$$CR = \frac{\frac{(3.14-3)}{(3-1)}}{0.58} = 0.12 > 0.1$$

Which shows that the matrix is inconsistent.

However, there is a question here; where does that limit of inconsistency of 10% come from?

Bozóki and Rapcsák (2008) define it as a rule of thumb. It responds to Saaty's intuition, although Vargas (1982) gave a statistical interpretation.

5.2.1.2 Determining Criteria Weights

Weights are found using the geometric mean, then:

$$w_1 = \sqrt[3]{1*2*3} = 1.82$$

$$w_2 = \sqrt[3]{0.5*1*0.5} = 0.65$$

$$w_3 = \sqrt[3]{0.33*2*1} = 0.87$$

Finding the ratios between weights
$w_1/w_2 = 1.82/0.65 = 2.80$ against a preference of 2 between C1 and C2
$w_1/w_3 = 1.82/0.87 = 2.09$ against a preference of 3 between C1 and C3 against 3
$w_3/w_2 = 0.65/0.87 = 0.75$ against a preference of 2 between C2 and C3.

Examining the above results, it is evident that the DM underestimated the importance of C1 on C2 in 40%, and overestimated that of C1 on C3 in 30%, as well as that of C2 on C3 in 62.5%.

That is Eq. (5.1) is not satisfied, denoting inconsistency, as previously shown by the CR.

It appears that the relationship between C1 and C3 is the most important to create inconsistency.

According to Tomashevskii (2015), the average error between preferences and weights can be found using formula (5.4).

$$\text{Mean Random Error} (\text{MRE}) = \sqrt{2\,CI} \tag{5.4}$$

In this example, then:

$$\text{MRE} = \sqrt{2\,CI} = \sqrt{2*0.07} = 0.37 \text{ or } 37\%$$

Table 5.3 Matrix corrected by the DM

	C1	C2	C3
C1	1	3	3
C2	0.33	1	0.5
C3	0.33	2	1

Consequently, and according to AHP procedure, DMs need to adjust their judgements, and then they change the preference between C1 and C2 from 2 to 3. Now it is C1 = 3 C2.

Then the matrix is as depicted in Table 5.3.

Then:

$C_1 = 3\,C_2$

$C_1 = 3\,C_3$

$C_3 = 2\,C_2$

Computing the criteria weights with the geometric mean:

$$w_1 = \sqrt[3]{1*3*3} = 2.08$$

$$w_2 = \sqrt[3]{0.33*1*0.5} = 0.55$$

$$w_3 = \sqrt[3]{0.33*0.5*1} = 0.55$$

and the ratios between weights:

$w_1/w_2 = 2.08/0.55 = 3.78$ against 3 as in criteria C1 and C2 preference

$w_1/w_3 = 2.08/0.55 = 3.78$ against 3 as in criteria C1 and C3 preference

$w_2/w_3 = 0.55/0.55 = 1.00$ against 2 as in criteria C2 and C3 preference

$$\lambda = 3.05$$

CR = 0.04 < 0.1 then the matrix is now considered consistent.

An analysis of the 'errors' of the DM in estimated preferences of importance shows the underestimated importance of C1 and C2, and C1 on C3 in a 26% in both, and that of C2 on C3 in a 100%.

Again, the relationship between C1 and C3 seems to be the main cause of inconsistency.

Now, in comparing the cases 5.2.1.2 and 5.2.1.3, it can be seen that when the DM changed only one point (preference of C1 on C2, from 2 to 3), the weights changed sharply, as C1 shifted from 1.82 to 2.08, that is, an increment of 14.2%; in C2, the decrement was 15.4%, and for C3, the decrement was 36.7%.

$$\text{MRE} = \sqrt{2\,CI} = \sqrt{2*0.025} = 0.22 \text{ or } 22\%$$

With these results, it is indeed difficult to agree with Saaty when he talks about 'small perturbations' (Saaty 2001).

This analysis illustrates the large influence that a change in the preferences may have in the criteria weights, and could indicate that a ranking of alternatives can change simply by changing a preference in one unit. That is, the ranking or output can be very sensible to a change in the preferences or input.

5.2.1.3 Experts Comments and Opinions on This Subject

Alonso and Lamata (2006): The intuitive meaning of the 10% rule is skipped by several authors.

These authors' opinion on this subject	Arguments, observations, judgements, remarks and opinions from MCDM experts, researchers and authors on this subject
Pair-wise comparisons have many drawbacks to be used for decision-making	Bozóki et al.: the major drawback of Saaty's inconsistency definition seems to be the 10% rule of thumb

5.2.2   The Pair-Wise Method in AHP Constructs Artificial Relationships

If we compare two criteria such as *environment* and *social action* with respect to the goal *improve wellbeing*, and give each independent values, from '0', say for instance 2, for the first and 4 for the second, then we are measuring them in an absolute scale, and meaning that the first has a larger value than the second.

However, if we say that the second is two times more important than the first, then we are establishing a false relative relationship between those criteria, which in reality does not exist.

5.2.2.1 Experts Comments and Opinions on This Subject

Kunsch (2012): The basic root of this anomaly (Rank Reversal) is recognized there as being the artificial interdependency created between alternatives because of the pair-wise comparisons.

Song and Kang (2016): In other words, there is an inconsistency between weights by the AHP and subjective weights.

Comparison

These authors' opinion on this subject	Arguments, observations, judgements, remarks and opinions from MCDM experts, researchers and authors on this subject
Pair-wise comparison generates artificial relationships between criteria and alternatives	There is artificial interdependency created between alternatives because of the pair-wise comparisons
	Schoner et al. This paper shows that there is a necessary correspondence between the manner in which criteria importance are interpreted and computed and the manner in which the weights of the options under each criterion are normalized. In general, if this relationship is ignored, then incorrect weights are generated for options under consideration regardless of whether new options are added or deleted.

5.2.3   Criteria Preferences Must Consider Alternatives

If somebody needs to make a trip between A and B, 1000 km apart, either using the aeroplane or a bus and compares travel time with cost, then how can a decision be made, saying for instance that travel time is more important than cost when this question is formulated *From the point of view of convenience, which is more important travel time or cost?*

This question is incorrect, since to make that comparison he needs to know the differences in travel time and in cost, thus, the question should be *From the point of view of convenience, which is more important, travel time or cost, considering that there is a difference of 400 Euros in favour of the bus, and a difference for 10 hours in favour of the airplane?*

In AHP the DMs don't have these values for the alternatives, consequently, it is impossible for them to pose this question.

5.2.3.1 Experts Comments and Opinions on This Subject
Salo and Hämäläinen (1997):
(a) The decision-makers need to understand that both the structure of the hierarchy and the criteria weights need to reflect the set of decision alternatives and their differences.
(b) Our analysis indicates that pair-wise comparisons should be understood in terms of preference differences between pairs of alternatives.
   Hulkower and Neatrour (2016): AHP is deeply flawed. The primary issue is that it relies strictly on pair-wise comparison and ignores critical information about the relationship to other alternatives.
   Dyer (1990a, b): He demonstrated that the original AHP may produce arbitrary rankings. He attributes the problem to AHP considering that the weights of the criteria do not depend on the alternatives (Triantaphyllou and Mann 1994).

Comparison

These authors' opinion of this subject	Arguments, observations, judgements, remarks and opinions from MCDM experts, researchers and authors on this subject
Pair-wise comparisons are meaningless if alternatives are not considered	The decision-makers need to understand that the criteria weights need to reflect the set of decision alternatives and their differences.
	Pair-wise comparisons should be understood in terms of preference differences between pairs of alternatives.
	AHP ignores critical information about the relationship to other alternatives.

5.2.4   The Ambiguity of Pair-Wise Comparisons in AHP

Pair-wise comparison appears to be an excellent procedure; however, it must be taken very carefully. If we have two different fruits as peaches and oranges, then we can't add them up, but they can be compared regarding a single feature they share, as is the satisfaction that they provide, based for instance in their respective sweetness. Then, a person can say that regarding sweetness, oranges are preferred.

Assume that a DM must make a pair-wise comparison between criteria *Environment* and *Social issues*. Even when they are not related, they can be compared, aiming for instance, to improving people wellbeing (the goal).

The DM compares them and determines that the first criterion is three times more important than the second. Don't consider for now, how he reached that figure, since there is another very important issue to be solved first, and it is linked with a fundamental question: On what attributes or characteristics of both criteria does he base his preference?

The question is necessary, because both criteria have many different meanings.

A set of projects may produce an improvement of the environment by creating more green spaces, by replanting, by eliminating mosquito infected pools or by recycling policies, that is, calls for maximization of benefits. At the same time, it can minimize environmental damage due to projects discharges into air, water and soil, or by erosion and so on.

Social issues may involve maximizing disposable income and providing adequate dwellings to people of limited means, and at the same time may call to minimize crime, poverty and number of kids on the streets.

Therefore, a question arises; when the DM compares environment with social issues to which of these aspects he refers to?

Probably, the answer of a practitioner using AHP will be: *It is near impossible to take into account all of this.* No, it is not impossible, but it is unfeasible to incorporate them in AHP, and because of that, the DM puts all the different aspects in the same bag.

This example shows the lack of realism of AHP.

AHP is based on psychological assumptions, however, none of that is needed; instead, the DMs must use their minds to think, to investigate and then to apply common sense.

5.2.4.1 Experts Comments and Opinions on This Subject

***Köksalan* et al.** (2013): The Marquis de Condorcet (whose name was Marie Jean Antoine Nicolas Caritat 1743–1794) produced several interesting results regarding holding fair elections. One of them is known as Condorcet's paradox, stating that majority preferences may be intransitive, even though individual preferences are perfectly transitive.

Salo and Hämäläinen (1997): AHP uses to elicit preference information about the alternatives that are typically of the form 'Which of the alternatives, Mercedes or Honda, is better with respect to quality and by how much?'. However, Watson and Freeling (1983), Belton and Dyer (1990a, b), among others, have argued that such value comparisons do not constitute an acceptable procedure of preference elicitation.

Kunsch (2012):

(a) The only way for eliminating this artificial interference between alternatives is avoiding pair-wise comparisons altogether. Instead, direct comparisons of alternatives must be made with independent anchoring values.

(b) A fundamental flaw is inherent to all pair-wise comparison methods, of which Rank Reversals are only one manifestation.

Hulkower and Neatrour (2016): Frequently, methods are chosen on an ad-hoc basis out of ignorance or because the outcome fits a preconceived notion of what should be correct.

Barakos (2019): AHP is not flawless but can be a very powerful tool if you can use it properly.

Hazelrigg (2019): AHP is not a mathematically rigorous method developed by Saaty.

Sehra **et al.** (2012): Inability of AHP is to deal with the imprecision and subjectivity in the pair-wise comparison process, which would improve in fuzzy AHP.

Saaty (1996): Not only does the importance of the criteria determine the importance of the alternatives as in a hierarchy, but also the importance of the alternatives themselves determines the importance of the criteria.

Comparison

These authors' opinion of this subject	Arguments, observations, judgements, remarks and opinions from MCDM experts, researchers and authors on this subject
Pair-wise comparisons is not an adequate method to be used for decision-making	Different authors regarding pair-wise comparisons articulate that:
	Artificial interdependency between alternatives,
	Avoid pair-wise comparisons,
	Inability of AHP in dealing with pair-wise comparisons,
	Preferences need to consider structure of hierarchy,
	Criteria weights need to reflect alternatives,
	AHP comparisons don't constitute an acceptable procedure,
	Pair-wise comparisons need to consider alternatives,
	A fundamental flaw is inherent to pair-wise comparison
	Individual preferences are mostly intransitives, not an acceptable procedure,
	There is artificial interdependency,
	Production of rank reversal,
	AHP is deeply flawed,
	AHP is not a mathematical rigorous method.
	Saaty: criteria determine the importance of the alternatives as alternatives determine importance of criteria,
	AHP is useful when there is not a utility function,
	AHP is not flawed,
	Belton and Stewart (2002): for instance, a respondent might state that pollutant concentrations are three times as important as costs. While the sentiment of this statement may make sense, it is completely useless for understanding values or for building a model of values.

5.2.5   Modelling Scenarios

For many AHP users, it appears that the size and the scope of the scenario is not important, and then they consider that the same method can be applied to any project.

This fact can be easily checked by a glance to the literature regarding the different kinds of projects 'solved' by AHP.

However, the most important aspect is that AHP does not have the capacity to tackle projects with characteristics that need to be inputted in the mathematical model, because the method has not been designed for that. This is not only an AHP glitch, because it is common in practically all MCDM methods. In general, nowadays, MCDM methods cannot model real-world scenarios, not even approximately.

Just as a simple example: suppose that a City Hall decides to give a good use to abandoned land in an island in front of downtown, which was a former commercial harbour and no longer in operation. There are several projects, such as (P1) construction of a bridge connecting the island with the mainland, (P2) construction of a new City Hall building, (P3) develop a technological centre and (P4) devote the area to construction of corporate offices and a park.

There are two heavy restrictions in this scheme: (1) The construction of the bridge is mandatory – it must be selected – and (2) It must obviously precede in time to any other construction, in order to supply materials, people and equipment during construction, as well as further access to whatever facility is selected, therefore, this precedence and its compulsory construction must both be inputted into the model. However, it must be subject to all criteria as the other projects, because its construction utilizes resources that are also used in the remaining projects.

If the bridge construction is not pre-selected, then the result may indicate to build any of the other undertakings without the bridge. It could be mathematically feasible, but definitively not fitting, since there would be no means to transport all material and heavy equipment needed. Can anybody assert where preferences enter in this very realistic and common scheme in the real world?

Of course, the scenario could be more complicated, as for instance if projects P2 and P3 are exclusive (meaning it is one or the other, but not both, for instance, because size of available land), while project P2 is inclusive regarding project P3 (meaning that both can be built). Obviously, there is need to input available funds, distances to a subway station, type of soil for each undertaking, inherent risks, water availability, volume of traffic on the bridge at peak hours and so on.

This project cannot be solved by AHP, because it considers relationships between these projects, however, AHP does not work with precedence or with mandatory partial results, let alone the effect of one alternative or criteria over another.

This example is indeed a non-complex scenario, nevertheless, nobody can seriously consider solving it with AHP or ANP.

Regarding relationships between alternatives and criteria in modelling, Saaty (2008a) says that *Not only does the importance of the criteria determine the importance of the alternatives as in a hierarchy, but also the importance of the alternatives themselves determines the importance of the criteria. Feedback enables us to factor the future into the present to determine what we have to do to attain.*

The first sentence is true, but not in the sense in Saaty's statement, and in addition, he does not explain the meaning of *feedback*, which makes difficult to agree with him.

In the same publication, he expresses that ANP considers dependency between the elements of the hierarchy and affirms that *Many decision problems cannot be structured hierarchically because they involve the interaction and dependence of higher-level elements in a hierarchy on lower-level elements. Therefore, ANP is represented by a network, rather than a hierarchy.*

This paragraph is interesting because it expresses that AHP cannot be used in certain scenarios. It mentions *dependence* from top-down and recognizing that some projects cannot be hierarchically structured because of dependency. Then, it appears that Saaty is explaining why ANP was developed using the network frame. This makes sense; however, he does not explain the meaning of dependence.

Dependency, according to Merriam-Webster Dictionary is *determined or conditioned by another*, and this is really the case in both ways. However, there is a difference:

(a) Criteria directly depend on alternatives, since the first are selected as a function of the second. Only when we know the alternatives, it is possible to define the criteria, since for instance, the set of criteria for a selection of sites for hydro-electric generation is not the same as in the case when the alternatives identify potential suppliers of raw material for an industrial plant.

(b) Alternatives selection certainly depends on criteria for evaluation, but not based on their relative importance derived from considering trade-offs, as AHP does. Alternatives do depend on the relative importance of criteria grounded on their capacity to select alternatives, and this capacity is a function of the dispersion of the respective performance values.

For instance, assume that three sites are pre-selected to generate electricity using dams. The three zones are in different geographical areas from alpine to desert landscape, and all of them are in independent rivers basins. Suppose that out of the many criteria the three sites are subject to, there are two that are pair-wise compared: *Flow of the rivers that will feed the lake behind the dam'*, and *'Impact on the wild-life.*

The DM rated the first as four times more important than the second, because according to his intuition, it is the primordial factor for power generation. It would be difficult to challenge this argument.

However, there is a problem of which the DM was not aware of, and it is that for whatever geophysical and ecological reasons, and using a 50-year statistic, there is evidence that the flow of each of the three rivers is practically the same. Consequently, this criterion, as important as it is, is useless, and practically may be eliminated, because its selecting capacity is near null, or expressed more technically, its quantity of information is too low to be used as an evaluator.

This amount of information can be quantitatively measured using the concept of *quantity of information*, or *entropy* (Zeleny 1974), and is based on a famous theorem that gave birth to the *Information Theory* as we know it today. That theorem was developed by Claude Shannon in 1948 (Shannon 1948).

Since the DM bases his preferences on intuition, he thought that the main criterion was the flow, ignoring that it is similar in the three rivers, and then, useless.

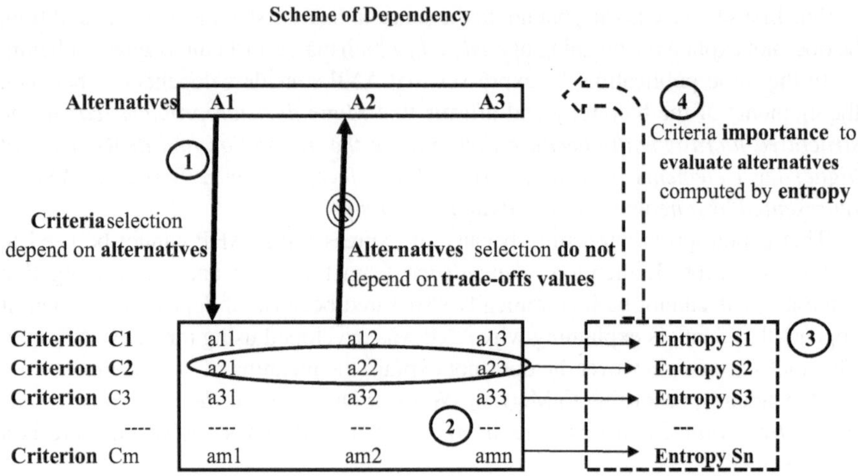

Fig. 5.1 Relationships between alternatives, performance values and entropy

Figure 5.1 depicts the scheme for the relationship between alternatives and criteria.

That is, given the alternatives, criteria are chosen according to them. When criteria are loaded with performance values, one for each alternative and each criterion, their respective quantity of information is used to select alternatives.

5.2.5.1 Experts Comments and Opinions on This Subject

Ksenija et al. (2015): Traditionally, MCDM methods only allow the establishment of linear dependence between criteria, so they only allow for simplified models that are mostly inadequate for modelling real-life problems.

Velasquez and Hester (2013): The method (AHP) has experienced interdependence problems between criteria and alternatives.

Carlsson et al. (2008): In modelling real-world problems (especially in management sciences), we often encounter MCDM problems with interdependent objectives.

Ishizaka and Labib (2009): The modelling effort generally defines the richness of the output.

Massachusetts Institute of Technology (MIT):

(a) The essence of project management is the model building approach, that is, an attempt to capture the most significant features of a decision by means of a mathematical abstraction.

(b) Models are simplified representations of the real world. They must provide a complete and realistic representation of the decision environment required to characterize the essence of the problem under study.

(c) Analytical models introduce the highest degree of simplification.

(d) Managers should elucidate the basic questions to be addressed by the model and then interpret the model results considering their own experience and intuition.

(e) Take as much as the right as possible in the model spectrum.

Comparison

These authors' opinion of this subject	Arguments, observations, judgements, remarks and opinions from MCDM experts, researchers and authors on this subject
Modelling must consider as much as possible all characteristics of the scenario	Interdependence problems,
	High degree of simplification,
	Must interpret the model results,
	(Koen 2008): some alternatives may have consequences attached to them. In modelling preferences, attention should be paid to assessing the consequences of the alternatives, if applicable.
	Some of the reasons motivating the development of different methods are due to specific requirements from the practical problem context.
	(Koen 2008): the use of a tree method has an advantage in that it provides more structure and a better overview of the classification method, but it also has its own drawbacks. For instance, it seems to operate on a very linear approach to the classification (you cannot move down two 'branches' at the same time, so a clear choice must be made at each node).

5.2.6 Report to Stakeholders and Lack of Rational Answers from DMs

If decision-making must be performed following intuitions, moods and perceptions of the DM, as AHP supporters assert, then it is in reality too daring or too naive. Granted, that this procedure is understandable and useful for both personal and corporate problems as commented, but definitely not applicable to even slightly complex scenarios.

For the authors, as engineers, and used to work in large problems, it is very difficult to accept this procedure; It is hard to imagine deciding by intuition that a diesel engine for a car is better that a gas engine, and deciding that the first is two times better than the second, and it is weird to conceive a financial officer in a company or an accountant, concluding that the Internal Rate of Return (IRR) is three or five times more important than the Net Present Value (NPV). Of course, the DM can have preferences, but quantifying this preference is something different, and this is one of the greatest and most criticized features in AHP.

Even if the DM oversees applying a methodology to select the best alternatives for a firm, the stakeholders are responsible for its approval or rejection. Therefore, and naturally, once a result or decision is taken by the DM, he will be questioned by stakeholders (SH), managers and directors, who can legitimately ask on what basis a result was attained, since for sure they are not going to blindly accept a DM conclusion.

Just as an example, when the DM uses AHP, is he in condition to answer these questions, often heard in the real world, and more important to justify his preferences?

Let's see:

Assume, for instance, this interchange between stakeholders (SH) and a DM when delivering his result:

SH – *Could you explain why you consider that Cost is more important than Benefits?*

DM – *Because intuition,* or *because I feel that Cost can be controlled better than Benefit.*

SH – *Therefore, you are deciding what is the best way to spend our money based on your feelings and intuitions. And, on what grounds do you base your preferences if you did not consider the different options or alternatives?*

DM – *Because that is AHP methodology.*

SH – *Have you analysed why a preference of one criterion over another should be constant?*

DM – *No, but the method assumes it.*

SH – *You know that our location project involves installing a car assembly plant in North America and that we have identified three potential cities, New York, Toronto and Miami, and your study selected Toronto.*

DM – *Well, yes, because this is the location with the largest score.*

SH – *Granted, but the problem is which are the assumptions and data you used to get that score. At the very beginning of your analysis, you were comparing two criteria, namely, availability of expert people in this kind of undertaking, and transport facilities, since about 60% of our production is exported to Europe, and 40% sold in the United States, Canada and Mexico, and you said that expertise is more important than transport. At the end, this preference, without your prior knowledge of course, favoured Toronto, because it's very well-known expertise in car manufacturing and closeness of steel mills.*

However, since you were not considering the alternatives for this comparison, did you not realize that Toronto harbour is closed for more than three months in Winter time because of ice, and then, taking into account this circumstance, transport is more important than expertise. You can see that ignoring the nature of the options produced a non-satisfactory decision, about which you can't give a solid explanation on which you based your judgement.

Had the DM considered this circumstance, most possible the result would have been different. As a bottom line, the DM can't make comparisons between criteria if ignoring the alternatives.

5.2.6.1 Experts Comments and Opinions on This Subject

DTLR- (UK Government) (2001):

(a) We are really considering the consequences of the options, not the options by themselves.

(b) Criteria are specific measurable objectives.

(c) Criteria express in many ways that options create value.

(d) Build a consequence table for each option.

(e) The MCDA must include criteria that are of the concern to all stakeholders.

Habenicht **et al.** (2002): The decision-maker must design a set of criteria that reflects the various consequences arising from the choice of an alternative

Koen (2008):

(a) Some alternatives may have consequences attached to them. In modelling preferences, attention should be paid to assessing the consequences of the alternatives, if applicable.

(b) Belton and Stewart indicate that in order to warrant a formal MCDA modelling or analysis procedure, a problem must have substantial consequences, have impacts that are long term or affect many people, and be of a nature in which mistakes may not easily be remedied.

(c) Decisions matter when the level of conflict between criteria, or of conflict between different stakeholders regarding what criteria are relevant and the importance of the criteria, assumes such importance that intuitive 'gut-feel' decision-making is no longer satisfactory.

Comparison

These authors' opinion of this subject	Arguments, observations, judgements, remarks and opinions from MCDM experts, researchers and authors on this subject
Report to stakeholders needs to answer their questions and give them valuable and useful information	Need to consider consequences
	Set of criteria reflecting consequences from choices
	Consequences may be substantial and may impact many people
	Gut feeling is not satisfactory

5.2.7 AHP Incapacity to Solve Complex Problems

Many authors and even consultants on MCDM affirm that AHP can solve complex problems. This is a misconception. This can be easily corroborated by examining the myriad of different projects 'solved' by AHP.

For many people, a complex project is the one with a large number of criteria and sub-criteria, which is not complex but perhaps industrious and time-consuming. Complex projects include many aspects that can't even be modelled in AHP, simply because it has not been designed for that.

It is proposed here a simple problem with an elemental hierarchy, and even in this case, it can't be solved by AHP.

Problem:

A domestic garbage incinerator must be built and there are two potential locations A and B, and then only one needs to be selected. In turn, each one of them may have two options in accordance with the height of chimneys (80 or 120 metres).

These four alternatives are subject to 12 criteria related to environment, distances, odours and electric supply.

However, there is a mix of positive and negative performance values, for instance, for criterion *Chronical exposure for population to NOx*, there is a value for each option; in site A, and with an 80 metres height chimney, it is −0.29, meaning that this option has a negative effect valued in 0.29 (normalized value), while for B and a height of 120 metres, it has a value of 0.10.

Another criterion is *Level of deposition of heavy metals on the ground* with a value of 0.17 for A and 0.15 for B with a 120 metres chimney.

How a DM can establish from the point of view of goal *Minimizing contamination,* that the first criterion is say three times more important than the second. On what grounds?

Well, he can say that population health is more important than soil contamination, and this may be true. However, he did not consider or perhaps did not know, that NOx can be dissipated into the air in hours, according to prevailing winds, while the second may remain for years on the soil and contaminate crops and aquifers. After knowing these facts, he may perhaps change his preferences.

At the same time, each criterion has two limits, a minimum and a maximum, and some criteria call for maximization while others for minimization.

How AHP could solve this simple problem?

It can't.

This is a real MCDM problem due to L. Tasca, Laura Magistrale, University of Torino, Italy, and yes, she solved it, albeit of course, not with AHP.

5.2.8 Criteria Independency

AHP assumes that criteria are independent, but what happens when they are related, as for instance safety and speed in a road? The higher the speed the lower the safety, that is, an inverse and probably non-linear relationship. This poses a very strong restriction in using AHP method, since most criteria are related directly or indirectly.

5.2.8.1 Experts Comments and Opinions on This Subject

Barba-Romero (Translated from Spanish) (2000):

(a) Criteria independence is very seldom seen in practice, and then it is absurd to pretend that it exists.

(b) This property is the most immediate to consider when modelling, when trying that all decision elements relevant to the problem be reflected as criteria.

Kasperczyk and Knickel (2006): Note that AHP, as all MAVT (Multiple Attribute Value Theory) methods, can only be applied when the mutual preferential independence axiom applies.

Improta et al. (2018):

(a) Among MCDM techniques, the AHP still suffers from some theoretical disputes. One major criticism is that the assumption of independence among the criteria can be considered a limitation of the AHP in certain cases.

(b) ... It (AHP) cannot be considered suitable for the modelling of complex problems, which are characterized by dependencies, interactions and feedback and especially by the dynamic nature of the decision taken.

(c) Another criticism of the AHP is the inherent static nature of the decision, which means that the method is ineffective in case of the future need for a medium/long-term decision.

Saaty (1994): In fact, one of the main aspects of the AHP is the assumption of independence between the various levels of the hierarchical structure in terms of both the criteria and sub-criteria.

Lee (2010): Since the criteria (factor) weights are traditionally computed by assuming that the factors are independent, it is possible that the weights computed by including the dependent relations could be different. Therefore, a hierarchical representation with a linear top-to-bottom structure is not suitable for complex system (Chung et al. 2005)

Öztürk (2006): In multi criteria decision making (MCDM) theory, the general assumption is to assume that the criteria are independent. This makes optimal MCDM solutions less useful than they could be and a decision-maker who accepts an optimal solution from the model cannot be sure that he has made the correct trade-offs among the objectives.

Yüksel and Deviren (2007). Although the analytic hierarchy process (AHP) technique removes these deficiencies (refers to SWOT), it does not allow for measurement of the possible dependencies among the factors. The AHP method assumes that the factors presented in the hierarchical structure are independent.

Comparison

These authors' opinion of this subject	Arguments, observations, judgements, remarks and opinions from MCDM experts, researchers and authors on this subject
AHP considers criteria independency when more often than not they are dependent	It is very seldom seen in practice,
	Absurd to pretend that it exists,
	AHP works only when independence axiom works,
	Saaty: a main aspect of AHP is assumption of independence at various levels,
	If weights are computed assuming dependence, then it is possible that they are different when considering independence,
	Hierarchy representation from top-down is not suitable for a complex system,
	A DM accepting a solution that considered independence cannot be sure that trade-offs are correct.

5.2.9   Quantifying Preferences

Putting a quantitative value when comparing abstract aspects is weird, to say the least.

For instance, establishing in a football game, that team A was 2.5 times better than team B, regarding quality, borders incongruity. The DM can, of course, express preferences on an issue, but quantifying it is utter non-sense. To say nothing if the DM has to put a value to the relation between two very subjective criteria such as 'happiness' and 'goodwill'.

Many MCDM professors put as an example – invented by Saaty – where there are several geometrical shapes and then asking students to estimate the relative areas of these figures. When this is done, results are compared with exact values and happily say *You can see how the guess approximate reality.*

More than one student rebutted the professor's assertion arguing that the comparison is biased, since they had a visual reference for making comparisons, which is inexistent when making comparisons in the real world, for instance, between environmental impact and manpower, when related to a main objective.

5.2.9.1   Experts Comments and Opinions on This Subject

Ishizaka and Labib (2009): Criticizes this approach since in ratio scales an absolute zero does not exist. This opinion is also shared by Barzilai (2001), Dodd et al. (1995).

   Saaty (1980): Advocates it as the best scale to represent weight ratios. However, the cited examples deal with objective measurable alternatives such as the areas of figures, whereas AHP mainly treats decision processes as subjective issues.

   Bernasconi **et al.** (2007): Direct measurement methodologies that rely on assigning a number to a stimulus because it corresponds to a number named in the instructions should be looked with incredulity (Narens 2006, p 298).

   Buchanan **et al.** (1998): In AHP, the ratio scale makes sense when dealing with distances, areas and so on but not when dealing with comfort, image or quality of life because there is not a reference level.

   Barzilai (2001): Performances can't be represented with ratio scales, because there is not an absolute zero.

   Dodd **et al.** (1995): Share Barzilai opinion.

   French (1983):

(a) The 1–9 scale has the potential to be internally inconsistent.
(b) The scale does not have a mathematical foundation.
(c) Weights are elicited before measurements (units) for criteria have been set.

Comparison

These authors' opinion of this subject	Arguments, observations, judgements, remarks and opinions from MCDM experts, researchers and authors on this subject
It does not make too much sense quantifying preferences following pre-established rules	In ratio scales, it does not exist an absolute zero
	Saaty: justify the method through an example where results are known and visual,
	AHP ratio scale makes sense dealing with distances or areas, but not with subjective issues,
	Performances cannot be represented with ratio scale, because there is no absolute zero,
	The 1–19 scale has the potential to be internally inconsistent,
	No mathematical foundation for the scale,
	Weights are elicited before measurement for criteria
	Song et al.: …third, since comparisons in AHP are made only using the scale of integers 1–9 and their reciprocal numbers, the proportionality between the importance values of factors is not always satisfied

5.2.10 Quantitative Data

In scientific international gatherings on MCDM and in a scientific forum, often practitioners ask about how to work in AHP with quantitative values, that is, with real and documented quantities, such as costs, performances, prices and so on and mixed with qualitative values. Not surprisingly, and these authors bear witness of this, they were given only void words and 'advices', but not a procedure, and then these people were unable to solve their problems. This practice seems to be very common in AHP defenders.

It appears that AHP theory commands that if you have reliable data from documented sources, you better forget them, because intuition is more important according to some AHP supporters.

5.2.11 The DM Should Be a Multiple Technical Expert

Since the DM has to compare the importance between criteria that may belong to very different and unrelated fields, AHP assumes that the DM has the enormous capacity of being at the same time an expert in economics, financing, purchasing, engineering, accountancy, environment, social issues, politics, agriculture and so on. As normally happens, in a scenario, there are a large variety of criteria related to most of these disciplines, and the DM must pair-wise compare them.

Naturally, when confronted with this issue, some people say that this is not the case, since normally there are various experts. True, in very large projects, but not in ordinary projects, where, however, many different fields may exist for criteria. This is another proof of the lack of reasonableness of AHP.

5.2.11.1   Experts Comments and Opinions on This Subject

Velasquez and Hester (2013): (AHP), due to the approach of pair-wise comparisons, can also be subject to inconsistencies in judgement and ranking criteria.

Konidari and Mavrakis (2007): (AHP) does not allow [individuals] to grade one instrument (policy measures) in isolation, but in comparison with the rest, without identifying weaknesses and strengths.

Comparison

These authors' opinion of this subject	Arguments, observations, judgements, remarks and opinions from MCDM experts, researchers and authors on this subject
The DM cannot be an expert in all fields	There could be inconsistencies in judgement and ranking criteria,
	AHP works by comparisons, not isolation, and then without identification of weaknesses and strengths.

5.2.12   The AHP Tries to Subordinate Reality to a Theory

Saaty advices to work with a number of criteria less than 9, although the number of sub-criteria can be larger. In other words, in a complex project where there are many different fields for criteria, may be involving economics, engineering, environmental restrictions, social issues, political, sustainability demands, water availability, land use restrictions, transportation and supply problems, efficiency, safety and so on as criteria, what is the solution to follow Saaty advice?

That is, you are advised to subordinate reality to an assumption.

5.2.12.1   Experts Comments and Opinions on This Subject

Cole (2018): Solutions to complex problems cannot be imposed upon the system. Instead, they emerge from the interactions that take place within the system.

Comparison

These authors' opinion of this subject	Arguments, observations, judgements, remarks and opinions from MCDM experts, researchers and authors on this subject
AHP conditions reality to a theory	Solutions cannot be imposed

5.2.13 Relative Importance Between Criteria Is Considered Constant

The AHP theory also takes for granted that once the relative importance between criteria is determined, they remain constant for each criterion, and that their unit values are valid to be used for choosing between **any pair** of alternatives. In fact, this may apply when comparing two alternatives, such as A and B, but perhaps does not hold when comparing an A and C pair.

It is odd that to evaluate importance of criteria, to further evaluate pairs of alternatives, it is done without considering them. AHP places alternatives at the bottom of the hierarchy, when they should be at the very beginning. The reason is simple: How can the DM establish the criteria to evaluate alternatives if he does not know them?

5.2.14 AHP's Fundamental Scale Has Limits 1 and 9

These authors have always wondered what happens when it is decided that criterion C1 is three times more important than C2, and C2 four times more important than C3, then, C1 is twelve times more important than C3, but the scale has a maximum of nine! How does the DM proceed in this case?

We have made this same question several times and are still waiting for an answer.

Saaty says in a paper that 9 is not a limit, and even that the scale can span from 1 to ∞. Well, at least it makes sense, but how is it inputted in the decision matrix, because there could be very different values in order of magnitude?

5.2.14.1 Experts Comments and Opinions on This Subject

Asadabadi **et al.** (2018):

(a) If A is very much more important than B, and B is much more important than C, then there is no number in the scale that can define the relations between A and C.

(b) If A is slightly more important than both B and C, using the AHP matrix (Table 5.4) the calculated importance weight of A becomes twice that of the weights of B and C, which is far different from being slightly more important.

That is, suppose a matrix like this:

There is not a perfect consistency in this matrix, since for perfect consistency it should be:

$a_{12} \times a_{23} = a_{13}$. That is, if C1 = 3 C2 and C2 = 4 C3, then C1 must be $3 \times 4 = 12$ C3, instead of 6 C3.

XLSTAT (2019), from the Web, addresses this issue, when the author states that:

Comparison matrix A is said to be consistent if $a_{ij} a_{jk} = a_{ik}$ for all i, j and k. However, we **shall not force the consistency. For example, B > A has 3 > 1** *value and A > C has value* **5 > 1**, *we shall not insist that B > C must have* **15 > 1** *value. This* **too much consistency is** **undesirable** *because we are dealing* **with human judgment.** *To be called consistent, the*

Table 5.4 Lack of transitivity or consistency

Criteria	C1	C2	C3
C1	1	3	6
C2		1	4
C3			1

rank can be transitive but the values of judgment are not necessarily forced to multiplication formula $\boxed{a_{ij} * a_{jk} = a_{ik}}$

Really, this explanation by XLSTAT, which according to their Web 'is the preferred tool for statistical analysis in businesses and universities', does not explain anything, starting with the statement about 'too much consistency', which according to its own definition, consistency is only one, and then trying to justify the high resulting value, resorting to psychological explanations.

Comparison

These authors' opinion of this subject	Arguments, observations, judgements, remarks and opinions from MCDM experts, researchers and authors on this subject
It appears that this table is not fit when the transitivity comparison reaches large values	If A is very much more important than B, and B is much more important than C, then there is no number in the scale that can define the relations between A and C
	There is no relation between the ratio in the table and the ratio between priorities
	Franek and Kresta (2014): results and their comparison show that judgement scales play a significant role in AHP decision-making.

5.2.15 A Preference and the Peculiar Meaning of Its Inverse Value

To better illustrate this point, consider the following example:

A person wants to buy a car; he has in mind what he needs or prefers in terms of comfort, speed, price and performance, say in that order.

There are normally several makes with these characteristics; therefore, it is possible to make comparisons between similar vehicles. If the person decides that for him comfort is eight times more important than speed, then does it mean that speed accounts only for 12.5% of his interest? Comfort might be very important for him because he spends many hours on the route, but does it mean that because of that he may settle for a car very comfortable, but with a maximum speed of 85 km/hr?

This shows another fallacy of AHP, since it equals a mathematical property to a real scenario.

5.2.16 Determination of Criteria Importance Must Contemplate the Alternatives

This is one for the authors' one of the most perplexing aspects. The DM may be earnestly working for producing a good scheme of his preferences. Trouble is, that his preferences are related to a global objective, for instance, to maximize company penetration in a market, and in so doing, he does not consider the alternatives that will be used to reach that goal.

Thus, from the point of view of getting his objective, the DM believes that criterion C1 (quality) is more important than criterion C2 (price). Fine, regarding what product?

It could be that alternative or product A4 has a lower quality but performs better than product A1, with higher quality.

AHP does not contemplate that situation since it assumes that quality is always more important than price. Thus, the DM is blinded because he has to select between two criteria without knowing to what they will be applied. The logical question would be: *From the point of view of market penetration, what is more important, price or quality, when comparing alternatives, A and B?* As seen, this reasoning brings us back to Sect. 5.2.13, regarding pair-wise comparisons as a constant.

5.2.16.1 Experts Comments and Opinions on This Subject

Perez et al. (2006): In general, these authors attribute this failure *(Rank Reversal)*, to the fact that, in Saaty's method, the weight of each criterion is independent of the evaluations of the available alternatives with respect to this criterion, which is caused by the way in which the method elicits these weights from decision-makers. Recently, Salo and Hämäläinen (1997) and others have expressed similar criticism of AHP.

Comparison

These authors' opinion of this subject	Arguments, observations, judgements, remarks and opinions from MCDM experts, researchers and authors on this subject
Criteria relative importance cannot be determined if alternatives are not considered	Rank reversal in AHP is attributable that weights for criteria are independent of alternatives.
	Song and Kang (2016): if the priorities are decided by the comparison of A, B and C, the importance values of A, B and C should be maintained, even if D is added for comparison, but since priorities among A, B and C are changed due to the addition of D, hence the reliability of the result is lowered.

5.2.17 The DM Preferences Do Not Consider the Real World, They Only Exist in His Own Universe

The DM, since he does not care for reality, is forced to 'fabricate' his own reality, that is, as he perceives it, in a scenario that is only in his mind, and which could be totally different to the real world. In other words, he is working in his own universe, tucked in his bubble, without any connection with the real world, and aiming to get transitivity. However, the real world is very intransitive and then normally inconsistent as Triantaphyllou (2000) expresses as *In the non-consistent case* (refers to matrices*), which is more common in real life applications....*

The misconception of this system can be appreciated if we consider two DMs. Each of them will work inside his own bubble, in their own perception of reality, and probably reaching a different result. However, reality is only one, and not two different, as they perceive it.

The beauty in using real-world examples is that they allow to appreciate that problems in the world are not as simple as AHP makes users believe, and that can be solved in a hierarchical structure and by preferences. Unfortunately, they are complicated, and for that reason, they have to be addressed as they are, not as we want them to be for sake of simplicity, and then, what we have to do is to consider as close as possible all real characteristics of a scenario and try to find a solution after these characteristics have been quantified according to real data, or from a reasonable analysis.

5.2.18 Where Is the Logic of the DM Correcting Himself Just for the Sake of Transitivity?

The DM looks for transitivity in his own previous preferences, and if he does not get it, it is forced by a mathematical formula to reassess his previous assessment. The AHP mentions that they are only small changes, however, it is not so.

He establishes, for instance, that criterion A is five times more important than criterion B, but he finds that his intransitivity is larger than 10%. So, he must abide to a formula and lower the importance in order to make ends meet, say to four times. That is, with respect to his original assessment, he decreases it in a $(5-4)/5 = 0.20$ or 20%, which is not precisely small.

Of course, corrections are normally needed, but when they come from a rational thinking and not from a formula. This is a good example of a human being subject to a machine.

In other words, he is forced to reach transitivity, even with a 10% tolerance, to justify that his preferences obey the transitivity requirement.

Where is the logic of this approach?

Probably some practitioners ask themselves why it is necessary to have transitivity. Wouldn't it be more rational to work with the values reached by the DM in his appreciation, regardless if there is transitivity or not?

We understand that this issue is related to the principle by which the eigenvector method guarantees an optimal solution only for a perfect transitive matrix.

5.2.18.1 Experts Comments and Opinions on This Subject

Fishburn (1991): Said *I cease to understand why transitivity should be a cornerstone of normative decision-making* and adds that *without transitivity axiom utility theory does not work.*

Asadabadi et al. (2018):

(a) When the number of elements to be compared increases, the ratio often falls beyond the threshold (0.1). This fact questions the credibility of the comparisons, so that such comparisons are returned to the decision-maker to improve. Usually, this process is repeated until the ratio becomes acceptable. In many applications of AHP, evaluators will start managing the numbers in order to decrease the ratio and satisfy the process while gradually paying less and less attention to what they really prefer (Asadabadi et al. 2018). Doing this may dramatically change the results.

(b) Ishizaka and Siraj (2018): Used three MCDM methods to compare several coffee shops and claimed that MCDM methods provide better rankings than intuitive approaches.

(c) It is a mathematically proven fact that when the order of a matrix increases to more than three, the inconsistency issue arises and increases exponentially as the number of the criteria and alternatives grows. If CR is more than 0.1, the user is blamed for providing inconsistent comparisons and the tables are returned to them for improvement.

(d) The problems with AHP become more serious when the decision-maker starts manipulating the value of pair-wise comparisons in order to get rid of inconsistency instead of performing a fair comparison between the elements.

Comparison

These authors' opinions of this subject	Arguments, observations, judgements, remarks and opinions from MCDM experts, researchers and authors on this subject
It is unreasonable that the DM must correct himself according to what a formula indicates	Why transitivity is a cornerstone of decision-making?
	Credibility of comparisons is questioned when the DM must go back in his estimates,
	This process provokes the DM paying less attention to what he is doing,
	Inconsistency arises exponentially with number of criteria,
	The DM manipulates the values to get rid of inconsistency in lieu of performing as fair comparison.

5.2.19 Normalization

AHP divides each value by the sum of all of them. That is, their normalized sum is 1. Now, if DMs decide that criterion C3 (price) must be increased for whatever reason, as is happens in sensitive analysis, they are taken for granted that criterion C4 (demand) will diminish linearly. However, the relationship between price and demand is not lineal, according to the economics theory.

5.2.19.1   Experts Comments and Opinions on This Subject
Ishizaka and Labib (2009): Main advantage of the ELECTRE method (over AHP) is that it avoids compensation between criteria, which distorts the original data.

   Checkland (1983): The divergence between theory and practice is no longer taken as a requiring proof; it is taken as given.

Comparison

This author opinion of this subject	Arguments, observations, judgements, remarks and opinions from MCDM experts, researchers and authors on this subject
AHP normalization to unity means that all weights and corresponding criteria are lineally interrelated, which may be not true	AHP needs compensation but this distorts original data
	There is large divergence between theory and practice; now is taken as given

5.2.20 The Use of the Eigenvector Method

Saaty decided to use the eigenvector (EV) to elicit priorities, which is unobjectionable.

   However, the EV gives exact values only up to 3 alternatives. What happens when the scenario has more than 3 alternatives, may be 8, 10 or 40?

5.2.20.1   Experts Comments and Opinions on This Subject
Bana e Costa and Vansnick (2008):
(a) The vector of priorities given by AHP does not respect the fundamental measurement condition.
(b) We consider that AHP has a fundamental serious weakness that makes its use as a decision support tool really problematic. It also applies to another method that derives priorities from a pair-wise comparison matrix.

   Ishizaka and Labib (2009): Priorities only make sense for consistent or near consistent matrices.

Comparison

These author opinions of this subject	Arguments, observations, judgements, remarks and opinions from MCDM experts, researchers and authors on this subject
Using the eigenvector method does not always guarantee that priorities will be optimal	The vector of priorities given by AHP dos does not respect the fundamental measurement condition.
	AHP has a serious weakness due to pair-wise comparisons.
	Priorities only make sense for consistent or near-consistent matrices.

5.2.21 Criteria Weights

AHP obtains a set of priorities that are trade-offs (Belton and Gear 1983), which simply indicate how much loss corresponds to a certain gain. Even Saaty recognizes that these are not weights when in (Saaty 2008a), he asserts: *To make trade-offs among the many objectives and many criteria, the judgments that are usually made in qualitative terms are expressed numerically.*

For instance, if we make a selection of two restaurants A and B based on quality and price, a trade-off shows our gain in quality when we sacrifice price, that is, we get better quality but also pay more.

Which is the main difference between trade-offs and weights?

Trade-offs are just a measure of effects due to preferences and are based on personal liking, intuitions, fondness and predilections, while criteria weights measure the relative importance of criteria and are based on the information contained within each criterion, and this is fundamental, taking into account the fact that the criteria weights can significantly affect the outcome of the decision-making process.

The error in using preferences as criteria weighting can be assessed considering that these depend on the DM, while objective criteria, for a given problem, are always constant and are based on the amount of information contained in the initial data. Koen (2008) agrees with this assertion when asserting that *Weights are in fact subjective expressions of trade-offs which could roughly be equated to an expression of the importance of one compared to another.*

There is another problem related to the number of sub-criteria considered to weigh a criterion.

The relative importance of criteria falls in the dispersion or discrimination of its attributes; therefore, the alternatives must be evaluated considering this importance, not based on their relative trade-offs respecting a main objective.

5.2.21.1   Experts Comments and Opinions on This Subject

Stanujkic **et al.** (2013): Changes in criteria weights may have impact on the ranking and also that the different rankings may obey to different criteria weights, as well as different aggregation methods and different normalization procedures.

Grupp and Schubert (2010): Weights should be chosen on the basis of shadow prices.

Triantaphyllou (2000):

(a) In decision-making, the weights assigned to the decision criteria attempt to represent the genuine importance of the criteria.
(b) The intuitive belief is that the criterion with the highest weight is the most critical (Winston 1991, p. 754). This may not always be true, and in some instances, the criterion with the lowest weight may be the most critical.
(c) The DM can make better decisions if he/she can determine how critical each criterion is.

Vujičić **et al.** (2016): Taking into account the fact that the criteria weights can significantly affect the outcome of the decision-making process, it is clear that special attention must be paid to the objectivity of criteria weighing, which unfortunately is not always present in solving practical problems.

Kazibudzki (2013):

(a) Intuitive decisions are not supported by data and documentation and may appear arbitrary.
(b) Right eigenvectors ability to generate weights is the principal criticism (and mentions several publications about it).

Schoner and Wedley (1989):

(a) Criteria weights have different interpretations and implications, and there is no consensus in their meaning.
(b) Criteria weights are not directly related to the discriminating power of the criteria in AHP, by the partial values.
(c) The achievement levels of the optimal alternatives are not necessarily related to the criteria weights, nor the criteria importance.

DTLR – (UK Government) (2001):

(a) Weighted average depends on the assumption of mutual independence of preferences.
(b) A measurement of judgements needs to specify how well an option meets the objectives expressed by the criteria.
(c) Is it possible in practice to measure how well an option performs on these criteria?
(d) Usually, better levels of preference lead to higher values scores.

Cinelli **et al.** (2014): In AHP, weights indicate trade-offs between the criteria.

Zanakis **et al.** (1998):
(a) Ranking differences derive from the process of weighting the criteria.
(b) Using the same weighting vector, the ranking order may vary depending on the method used and as number of alternatives increases.

Yeh (2002):
(a) Proposed select a method using Shannon entropy via sensitivity analysis of attributes weighting.
(b) Also proposed using entropy for a measure for the degree of consistency of rankings.

Moshkovich **et al.** (2012): Found that significant differences in rankings are produced by criteria weighting and scale transformation, especially in the aggregation process of preference information.

Zardari **et al.** (2015):
(a) Pair-wise comparison has been often criticized for simply asking for the relative importance of criteria without reference to the scales on which the criteria are measured.
(b) Decision-makers to choose weights directly do not guarantee that these are valid.
(c) Main difficulty is to reconcile the inevitable inconsistency of preferences in real-world applications.
(d) In many real-world applications, human pair-wise comparison is highly ambiguous and uncertain.
(e) Decisions-makers are likely to be tired and lose patience deriving this process and therefore, may not make their judgement consciously.
(f) They may change their minds frequently in order to ascertain the acceptance of the CR.

Belton and Gear (1983): While the Saaty method is inherently powerful as a questioning device and easy to use, there is a degree of imprecision in the specification of what factors should be considered when determining the weight. It appears that in certain circumstances the method can give anomalous results arising from a misunderstanding of what is required in specifying the inputs.

Ma **et al.** (1999): Weights determined by subjective approaches reflect the subjective judgement or intuition of the DM, but analytical results or rankings of alternatives based on the weights can be influenced by the DM due to his/her lack of knowledge or experience.

Farkas (2007): If a Pair-wise Comparison Matrices (PCM) is inconsistent, even in the slightest degree, then the principal eigenvector components do not give the true relative dominance of the alternatives.

Comparison

These authors' opinions of this subject	Arguments, observations, judgements, remarks and opinions from MCDM experts, researchers and authors on this subject
AHP determine trade-offs between criteria and then assumes, without any mathematical foundation, that they represent weights, and even if they were, they are not suitable to evaluate alternatives	Weights should be based on shadow prices,
	Weights attempt to represent criteria importance,
	It is intuitive to believe that the highest way is the most critical; this is not always true,
	Objectivity of criteria weighting is not always present,
	Intuitive decisions not supported by data may appear arbitrary,
	There are different interpretations of criteria weights,
	The thinking that eigenvectors can generate weights is very much criticized,
	There is no consensus about what criteria weights are,
	Criteria weights are not directly related to the discriminating power of criteria,
	The scores of optimal alternatives are not necessarily related to criteria weights or criteria importance,
	Weighted average depends on assuming independence of criteria,
	Is it possible in practice to measure how well an option performs on these criteria?
	Criteria weights are in reality trade-offs,
	In AHP using the same weighting vector, the ranking may vary,
	Proposing using Shannon entropy (objective weights) methods for weighting criteria,
	Significant differences in rankings due to criteria ranking and scale,
	Pair-wise comparison is criticized because there are no references to the scale of measurement for criteria,
	There is no guarantee of validity for weights elicited by the DM,
	The real world is inconsistent,
	In many real-world applications, human pair-wise comparison is highly ambiguous and uncertain,
	There is the possibility of tiredness and loss of good judgement in the DM because of abundant comparisons,
	DM may change his/her mind in order to get a good CR,
	There is imprecision in determining weights
	Alternatives selection can be affected by the DM lack of experience,
	If the DM is inconsistent, even in the slightest degree, the eigenvector result is not reliable,
	If the value tree shows quality of life and economic measures as the two main criteria, but quality of life is evaluated only in terms of 2 sub-criteria, while economic measures are evaluated by 10 sub-criteria in 3 different levels of the tree, one may want to investigate the appropriateness of the quality of life sub-criteria.

5.2.22 A Project Is a System; Its Disaggregation Is Not a Good Practice

Normally a project is a system, because it involves relationships between several components, consequently, modifying one of them influences others. AHP ignores this feature when using different branches of a hierarchy and analyses and computes them separately.

Breaking down a project is a very good method to analyse and understand it, but not for solving. If three projects A, B and C are subject to a set of criteria, and only A and B are considered, this ranking can be obtained: B > A.

However, if the three alternatives are considered simultaneously, probably the ranking will be different. This simple experiment was proposed in the 1990s by Triantaphyllou.

5.2.22.1 Experts Comments and Opinions on This Subject

DTLR – (UK Government) (2001): Disaggregating is good for thinking but not to take a decision.

Ishizaka and Labib (2009): Some criteria have to be maximized and other minimized.

As a result, the output is a partial ranking.

Hobbs et al. (2003): An MCDM approach helps analysts and decision-makers think systematically about the problem.

Uniersia (2009): When a problem cannot be decomposed into independent facets, for example, then a model that requires criteria independence such as the AHP should not be applied.

Hulkower and Neatrour (2016): … 'Rather than solving the problem as a collection of parts, the approach must coordinate interactions among the different faces' (Saari 2010, p. 081006–7).

Comparison

These authors' opinion of this subject	Arguments, observations, judgements, remarks and opinions from MCDM experts, researchers and authors on this subject
A project is a system and as that, it cannot be broken down in parts and calculated separately	Disaggregation is good but not for decision-making,
	A problem cannot be disaggregated if its components require dependency,
	Better than solving different parts of a problem, it is better to coordinate their interactions

5.2.23 Sensitivity Analysis (SA) and the Way It Is Performed Using AHP

To perform a SA, AHP, as well as most MCDM methods, proceeds by progressively varying the importance of a criterion (input), to analyse how it impacts the ranking (output).

This procedure, in principle, is correct when using objective weights for criteria, but it is procedurally wrong, when using AHP derived trade-offs values, the so-called criteria weights (wi), and for several reasons, as follows:

(a) Only one criterion is selected to vary, and this choice falls on the criterion that has the maximum wi; this is an arbitrary procedure because that choice does not have any mathematical support. It can easily be proved that this concept is not always true and, in many cases, the most important criterion is one with an intermediate wi value or even that with the lowest value.

(b) A ranking is selected due to the compliance of many criteria. AHP is unable to identify which, out of the total, are the criteria involved, and which don't have any participation.

(c) As mentioned, normally, only one criterion is selected, in what is known as *One at a Time* (OAT) method, when in reality, all intervening criteria should be considered simultaneously, according to the *All at a Time* (AAT) method.

Since normally a ranking depends on several criteria – which again, need to be identified – it does not make sense, and it is erroneous, to use only one. Consequently, it is necessary to ponder the simultaneous variation of all identified criteria. In this way, the output will reflex the joint action of all intervening criteria, as happens in real-world scenarios.

As an example, assume a scenario for industrial location having, say, four sites or alternatives A1, A2, A3 and A4, subject to seven criteria such as, for instance: C1 (cost), C2 (internal rate of return or IRR), C3 (environmental damage), C4 (disposable income), C5 (available water), C6 (available land), C7 (available manpower) and the resulting ranking is: A4 > A1 > A3 > A2.

For each of these alternatives, there are certain criteria that define its selection. Say, for instance, that the best alternative is A4. For it, the identified criteria that conform to this alternative are C2, C3 and C7. The other criteria, even when they participated in this selection, are not significant.

Consequently, it is necessary to vary each criterion separately; however, their relative influence on the output or ranking must be considered simultaneously, not individually. In so doing, the procedure replicates the real world, because a price, for instance, may fluctuate due to different parameters, such as demand, competition, market share and so on, but the drive of its deviation is usually not related to the variation of only one of them, but to the changes of all of them considered together.

In reality, the procedure is more complex because each criterion has an allowed range for change, which is normally different from one to another, and this has to be contemplated.

That is, when the DM increments the importance of say C2, the allowed range of variation of criteria C3 and C7 changes, because there is a linear combination between the three selected criteria.

The AHP method considers that a change of a criterion always produces the same change, and this is right, as long as the change is within the allowable range, but ceases to be correct when the increase surpasses the upper or lower level of the range.

As can be seen, it is a rather complicated problem; however, AHP limits to perform SA by showing in a graph when an alternative is replaced by another, due to the variation of a single criterion, chosen arbitrarily. Naturally, the ranking changes and to analyse these changes is the objective of SA. However, these changes do not obey to a mathematical analysis but to a subjective decision, and consequently, the result may be erroneous.

5.2.23.1 Experts Comments and Opinions on This Subject

Triantaphyllou (2000): Intuitively, one may think that the most critical criterion is the criterion which corresponds to the highest weight. However, this notion of criticality may be misleading.

Comparison

These authors' opinions of this subject	Arguments, observations, judgements, remarks and opinions from MCDM experts, researchers and authors on this subject
Sensitivity analysis is not properly performed in AHP, because it uses trade-offs, which are not acceptable to register criteria importance variations	Using a criterion with the highest weight may be misleading

5.2.24 Rank Reversal (RR) in AHP

Given a problem, and obtained its ranking, RR is the change produced in this ranking when an irrelevant alternative is added or deleted.

This is one of the first problems found in the 1980s in AHP, although it is also common in other MCDM methods. At present, the reason for this happening has not been yet discovered, but it is a major drawback of AHP. Apparently, and according to some researchers, ANP does not produce this phenomenon.

Since it is discovered in 1983 by Belton and Gear (1983), RR has received a good deal of attention materialized by many papers trying to explain it, and others offering solutions generally in form of incorporating new concepts in the AHP mathematics.

In AHP, the performance factors of the newly aggregated alternative don't have any effect on the existing set of criteria, because their priorities have been estimated without consideration of the alternatives.

However, the incorporation of a new alternative or a deletion of an existing one modifies the initial decision matrix. The vector of the new alternative may indeed produce a unique change in the ranking because it could be better than other alternatives, but in this case, it should not produce a change in the existing preferences. Thus, this is not RR but the consequence of modifying the original matrix.

To understand this concept, assume that a person wants to rent an apartment and the realtor has shown him three, as A, B and C, and subject to different criteria such as location, price, neighbourhood, distance to the subway and so on. Assume that the client applies an MCDM method and he gets this ranking: Apart. B > Apart. A > Apart. C, therefore he is very inclined to rent apartment B. However, the vendor shows him apartment D; the client runs the MCDM again and now the ranking is Apart. D > Apart. B > Apart. A > Apart. C.

This is fine, there is no RR, because the addition has not changed the original preferences; it shows a new one, but the relational sequence existing in the original holds.

However, in many problems, it can be seen that the new ranking obtained by an MCDM method produces changes between the existent alternatives, and this is rank reversal. That is, if the original ranking was Apart. B > Apart. A > Apart. C, and another alternative D is added, and the software run, the new ranking may have Apart D in any position, for instance: A > D > C > B.

Observe that now, the original preferences have changed: Now A > B, and C > B. This is rank reversal, because there is no reason for these changes. This is what Belton et al. detected in AHP, and that was further also found in almost all MCDM methods.

There have been many mathematical explanations of this phenomenon but apparently, none of them is satisfactory. In the opinion of these authors, RR happens in AHP for two reasons:

1. Alternatives are considered in a ratio scale. Consequently, adding an alternative or deleting one may alter the precedence, because the interactions with the new alternative or the lack of interaction in case of a deletion.
2. Criteria weights are not changed. As a result, the new alternative is judged by a criterion which importance has been considered constant, when in reality its importance depends on the alternatives.

Some researchers have developed MCDM methods where each alternative is given a score, which is independent of the scores of the other alternatives; according to these scientists, this avoids rank reversal.

5.2.24.1 Experts Comments and Opinions on This Subject

Pérez (1995): …In general, these authors (Watson and Freeling, Belton and Gear, Dyer) attribute this failure (RR) to the fact that, in Saaty's method, the weight of each criterion is independent of the evaluations of the available alternatives with respect to this criterion, which is caused by the manner in which the decision-makers elicit the weights, through questions like: *How much important is criterion C1 than criterion C4, regarding the same goal?* Salo and Hämäläinen (1997), Weber (1997) and others have expressed a similar criticism of AHP.

Pérez et al. (2006): In this paper, we show another feature of AHP which may be, and in many application, contexts will be, an even stronger shortcoming of the method.

> *It consists in the fact that the addition of indifferent criteria (for which all alternatives perform equally) causes a significant alteration of the aggregated priorities of alternatives, with important consequences. In hierarchies with four or more levels, rank reversal may happen. Since in almost all applications of AHP the set of criteria is not fixed ex-ante but is variable and is constructed in accordance with reasons of relevance and simplicity, almost all applications of AHP are potentially flawed.*

Salo and Hämäläinen (1997): Despite claims to the contrary, the super matrix technique does not eliminate rank reversals.

(a) For Belton and Dyer, the problem of rank reversal is due to the lack of theoretical foundation, so they do not consider the AHP method acceptable.
(b) In response to these authors, Harker and Vargas (1990) argue that rank reversal is due to the misuse of a lack of theory and axiomatic basis.
(c) Saaty made an extension of the AHP method and proposed two methods to solve the problem, neither of which worked.
(d) …. This means that a DM's preference ordering between two alternatives changes when an alternative is added or removed, and this clearly contradicts the Principle of Independence from irrelevant alternatives.

Forman (1993):

(a) A consequence of AHP not satisfying the independence of irrelevant alternatives assumption of MAUT is to allow the real-world phenomenon of 'rank reversals' to occur.
(b) A technique that does not allow rank reversals is flawed.

Comparison

These authors' opinions of this subject	Arguments, observations, judgements, remarks and opinions from MCDM experts, researchers and authors on this subject
AHP produces rank reversal (RR) when adding or deleting alternatives, and this is unacceptable, and in these authors' opinion, this is due to using a ratio scale in the alternatives.	The RR is attributable to the fact that in Saaty's method it does not take into account the alternatives because of the way questions are formulated,
	Addition of an indifferent criterion causes significant alteration of the priorities of the alternatives,
	Because criteria are selected considering convenience and simplicity, all AHP applications are potentially flawed,
	RR is due to lack of theoretical foundation, and then AHP is not acceptable
	RR is due to misuse of theory,
	When ranking between two alternatives changes because of RR, it contradicts the principle of independence from irrelevant alternatives,
	Some authors attribute this failure to the fact that the weight of each criterion is independent of the evaluation, of the available alternatives with respect to this criterion,
	An addition of indifferent criteria causes a significant alteration of the aggregated priorities of alternatives, with important consequences,
	The super matrix does not eliminate rank reversal,
	It is due to theoretical foundations,
	Rank reversal is due to the misuse of a lack of theory and axiomatic basis,
	Saaty tried to solve it proposing two methods that never worked.

5.2.25 Time Dependency in a Portfolio of Projects

The process of selecting an alternative usually requires some DM time. In that period, it is usual that the original situation changes, and then new alternatives and criteria are added or deleted. For instance, assume that the DM is analysing diverse government policies, subject to a set of criteria, and then, when he has finished the pair-wise comparison, the government decides to add a new one policy.

The AHP method is simple to understand and easy to perform for small projects, that is with a few alternatives and criteria, and static. However, if for whatever circumstances data change, for instance, a new criterion or alternative is added or an existent deleted, it means to redo the whole process, and this can be very cumbersome and expensive. This is repeated when the sensitivity analysis is performed, since the DM may realize that a certain vital criterion needs to be added.

AHP in some way does consider time dependency, but, as Hashemkhani et al. (2016) state: *The model (AHP) is developed based on previous experiences from the past and did not say anything from the future.*

In this sense, AHP is superior to other MCDM models because it takes into account experience, while the others don't.

5.2.25.1 Experts Comments and Opinions on This Subject

Saaty (2008a, b): The importance of time dependency and its significance in real-world situations shall not be ignored either.

Lin (2008): DMCDM (Dynamic MCDM) is focused on topics which are performed based on information at different time periods.

Comparison

These authors' opinions of this subject	Arguments, observations, judgements, remarks and opinions from MCDM experts, researchers and authors on this subject
AHP assumes that all potential alternatives are executed at the same time, and that is not usually the case in a portfolio of projects, where they can start and finish at different periods	The importance of time dependency and its significance in real-world situations shall not be ignored either (Saaty),
	AHP works with experience from the past and ignoring the future,
	MCDM focuses on projects at different times.
	Ineffective in medium/long-term decision.

5.2.26 Future Events (Limiters) Need To Be Considered

Sensitivity analysis must consider the effect that exogenous aspects may have in the selection.

For instance, assume that there are four products or alternatives that are considered by export such as A, B, C and D, and that D has been selected as the best.

Once selected, this product or alternative may be affected by exogenous variables such as international prices, competition quality, world-wide demand and so on. Each one of these may vary differently and in a different sense, some increasing while others decreasing. It is necessary to compute the effect of these exogenous variables acting simultaneously and posing risks for the selected product. These risks need to be evaluated quantitatively to assess the threat to the selected product.

As far as this author information, there is not a MCDM method that can accomplish this objective.

5.2.26.1   Experts Comments and Opinions on This Subject

Hashemkhani **et al.** (2016): If we consider values, alternatives and criteria won't change that much but still some topics and actions may happen and effect on the decision-making process and priorities. These probable important actions and problems are called limiters in this research and model. Each of them or a combination between them can happen and affect the whole process.

Comparison

These authors' opinions of this subject	Arguments, observations, judgements, remarks and opinions from MCDM experts, researchers and authors on this subject
AHP does not allow for examining future outcomes on projects selected	AHP does not consider limiters
	These important actions or a combination between them can happen and affect the whole process

5.2.27 Not All Problems May Be Analysed Assuming That There Is Always a Hierarchy

The MCDM process may be defined as finding the best alternative when a set of them is subject to a set of criteria that are as well-defined objectives.

The AHP structure considers three main levels: the main objective, the criteria (and sub-criteria) and the alternatives.

Then, it analyses each pair of criteria regarding this main objective and getting trade-off values. Consequently, it ignores the cardinal definition of MCDM, since the only objective it refers to is the main objective and then not considering that all criteria are objectives by themselves, as it is a well-known fact.

5.2.27.1   Experts Comments and Opinions on This Subject

Öztürk (2006): Many decision problems cannot be structured hierarchically because they involve the interaction and dependence of higher-level elements on lower-level elements.

Song and Kang (2016):

(a) However, studies on the problem caused by the hierarchy structure itself are rare,

(b) In other words, this shows that the AHP has a weakness in hierarchy structure,

(c) As seen in the aforementioned example, the problem of a hierarchy structure is that values are varied depending on how the attributes are grouped, and the values of sub-attributes are changed depending on that of their upper attribute,

(d) The AHP has the disadvantages that values are varied depending on the shape of the hierarchy structure, as well as the difficulty in maintaining consistency.

Comparison

These authors' opinions of this subject	Arguments, observations, judgements, remarks and opinions from MCDM experts, researchers and authors on this subject
Because its hierarchical structure AHP is unable to represent most real-world problems	Hierarchy cannot be used in many scenarios because special interactions may exist,
	Cilliers (2001): they have (complex systemsl) to be dynamic and adaptable, not rigid or invariable. Consequently, the notion of hierarchy is resisted. In terms of the structure of organizations, it is often argued to that extent that there should be no hierarchies at all, they should be shallow and loose. There must be enough space for innovation,
	AHP has a weakness in the hierarchy structure.

5.2.28 One Criterion May Place a Restriction on the Optimal Achievement of Another Criterion

In some scenarios, some performance values may be dependent on others, and the same applies to resources. For instance, in selected places for building an airport, the number of flights may be improved if purchasing *Carbon Emissions Bonds*, and with different needs for each site.

This is a trading where a company, region or a country with low carbon emissions can sell to another party the right to use its surplus. Also, a certain resource may establish a limit to another resource. For instance, in a housing development, there is normally a criterion placing a limit to household water consumption. However, another restriction may put a limit to the total volume that can be extracted from a nearby lake.

Usually, there are environmental limits for discharges of noxious gases into the air, but this limit may be not constant because it depends on the ecological and wind conditions of each site, and expressed by another criterion.

AHP is unable to handle this type of scenarios.

5.2.29 Problem Structuring

According to Belton and Stewart (2010) seminal paper, a structured multicriteria problem is one for which alternatives and relevant criteria have been defined.

Why is it important?

Because the same authors emphasize that to mistake the problem involved is to cause subsequent inquiry to be irrelevant or go astray, and add that the final sentence identifies what the statistician Kimball (1957) labels as an error of the third kind or solving the wrong problem.

How is normally the genesis of an MCDM scenario?

Which can be first, the alternatives or the criteria?

All AHP problems follow the same hierarchical structure, however, not all problems in real world are structured similarly, and this lack of flexibility badly limits the use of the method. Its structure is elemental, good perhaps for trivial problems but not adequate to represent a complex scenario.

For instance, assume that there is an international call for tenders for selecting suppliers for steam turbines and steam boilers, that is, it is a set of compounded equipment and normally manufactured by different companies.

As is normal, joint-venture is permitted, that is, two different suppliers of different equipment, although bidding separately, are joining efforts and expertise to supply a set, however, each of them will be evaluated separately. Therefore, if one is selected, the other has also to be selected, which implies that both must get the same score. There is here a conditional aspect that says that both suppliers have the same score IF and only IF one of them is selected.

Look at this other example. There is a portfolio of projects A, B and C, but there is no interest in ranking them, since whatever the choosing, only one will be built. Consequently, the result is select or not select, and this means that the result has to be binary, that is '1' or '0'.

Many scenarios involve several different projects, all of them must be built, but not at the same time. Their execution depends on different aspects such as funds availability, technical reasons, precedence and so on.

It is evident that none of the three scenarios can be modelled in a hierarchical structure, and as Franek and Kresta (2014) assert: a different structure may lead to a different final ranking.

5.2.29.1 Experts Comments and Opinions on This Subject

Lee (2010): …. therefore, a hierarchical representation with a linear top-to-bottom structure is not suitable for complex system (Chung et al. 2005).

Belton and Stewart (2010), citing Keeney (1992):

(a) Invariably, existing methodologies are applied to decision problems once they are structured…. such methodologies are not very useful for the ill-defined decision problems where one is in a major quandary about what to do or even what can possible be done.
(b) A structured multicriteria problem is one for which alternatives and relevant criteria have been defined.

Comparison

These authors' opinions of this subject	Arguments, observations, judgements, remarks and opinions from MCDM experts, researchers and authors on this subject
AHP uses a structure that is adequate perhaps for trivial problems but not apt to model complex scenarios. This is attributable to the fact that scenarios not always follow a vertical structure, and even in the case of ANP, which is based on multiples connections, can handle this type of problems because in their structure there is no room to input certain characteristics that require more than establishing links	A linear top-to-bottom structure is not suitable for complex systems,
	...Such (MCDM) methodologies are not very useful for the ill-defined decision problems where one is in a major quandary about what to do or even what can possibly be done,
	Alternatives and relevant criteria must be well defined.

5.2.30 AHP Theory Regarding That Criteria Impact and Modify Another Criterion

Apparently, AHP considers that relationship and dependency are the same thing, however, they are not. The first simply indicates a link that could be weak or strong, while the second implies influence of one on another. We can say that for a manufacturing company, two criteria such as benefits and stock of raw materials are related, because normally stocking costs have an incidence on benefits, but it does not mean that it directly indicates in what proportion an increase of stocked raw materials affects benefits, since there are many other aspects that affect benefits, and may be stocks have most importance.

We consider that there is dependency when the performance values of a criterion may depend or are affected by variation of the performance values in another criterion.

This dependency can be captured in part by *correlation*; however, correlation does not necessarily mean *a cause and effect* relationship.

This notion of dependency is quite normal in most projects, especially when selecting alternatives subject to a set of criteria, and where some of them are correlated. Consequently, in many cases, these correlations need to be considered when examining the future behaviour of the best alternative. Examining many cases, it appears that correlation appears frequently, however, it is also investigating the degree of correlation, since for a correlation coefficient of a low value, say for instance 0.25, the correlation is weak and possibly it is not worth to consider it.

Assume as an example that a company wants to erect a wind farm to generate electricity and has three potential locations A, B and C.

The three locations have been selected among others considering the wind regime, that is, its speed, speed steadfastness, minimum and maximum speed, frequency of high temporal speed variations, turbulence by existing features such as mountains and so on.

A very important factor is wind speed distribution and considering that the energy produced depends or is correlated to or depends on the wind speed. This dependency can be sketched in a diagram with output in MW in ordinates, and wind speed in m/s in abscissa. Figure 5.2.

As can be seen, output increases non-lineally with wind speed, up to a maximum value. This maximum value may be linked to protecting the turbine blades for very strong winds, and then, by shutting down the equipment, which is a normal practice.

It can be seen that this is a complex scenario, and an aspect that should be considered in the site selection is the frequency of high wind speeds and high temporal variations that produce the shutdown of the turbine.

Consequently, in this case, it is necessary to compute this curve for each place. If for site C, there is a very high frequency of dangerous gusts of wind that can endanger the installation, then this has to be considered in the study, because it could very well be that the best site corresponds to a high frequency of shutdowns. Turning its idle, it's not convenient from the economical point of view.

All these conditions must be inputted in the model, probably using the *IF*.... *then*... proposition statement.

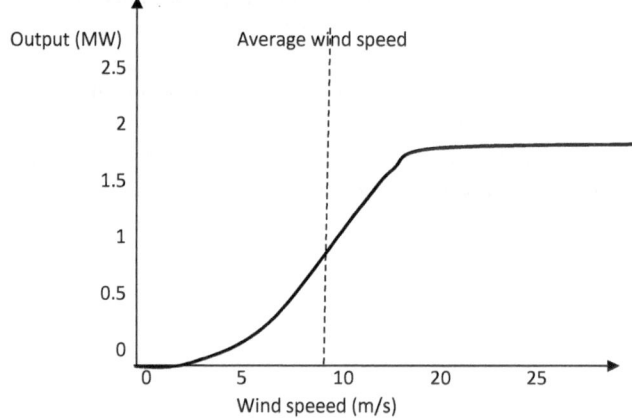

Fig. 5.2 Power output as a function of wind speed

5.2.30.1 Experts Comments and Opinions on This Subject

Liu et al. (2014): However, besides outer-dependence, correlation is another common effect between criteria which can be accounted for neither by the AHP nor by the analytic network process (ANP).

Perez **et al.** (2006): These authors also detected that *the addition of indifferent criteria (for which all alternatives perform equally) causes a significant alteration of the aggregated priorities of alternatives, with important consequences.*

Comparison

These authors' opinions of this subject	Arguments, observations, judgements, remarks and opinions from MCDM experts, researchers and authors on this subject
AHP apparently considers that dependency and impact are the same mechanism, and from there, it develops a theory using 'feedbacks' which is unrealistic and does not explain its meaning and uses.	Correlation is another common effect between criteria which cannot be accounted for neither by the AHP nor by the analytic network process (ANP),
It tries to justify that ANP can consider impacts, which is erroneous, and lead the user to false assumptions. The fact that neither AHP nor ANP can consider correlation is enough proof that this assumed characteristic does not exist	The addition of indifferent criteria can cause alteration in the ranking of alternatives.

5.3 Conclusion of This Chapter

As posted repeatedly, this book was written with the purpose of enumerating and analysing critically, the AHP method and to demonstrate, not mathematically, that it is unable to solve problems in scenarios even with a minimum level of complexity. A rough definition of complex problems from the point of view of MCDM was also proposed, explained and illustrated with several examples.

Further, 30 AHP shortcomings detected and analysed by these authors were included and illustrated through examples. Independently, a research took place on these same shortcomings, considering hundreds of published papers, books, the Web, diverse examples, university reports and so on. Its objective was to find coincidences or lack of them for these shortcomings and then be able to reach a conclusion. Table 7.1 in Chap. 7 condensates the opinions of more than 90 researchers.

In this table, it may be noticed that there are no mentions from experts, corresponding to some particular analysis performed by the authors. This lack of information is due to the fact that some shortcomings are product of the research of the authors of this book and apparently were not detected before.

For this reason, out of 30 shortcomings, there are 25 coincidences, that is, the same number of subjects in where researchers and our opinions can be compared, and another 5 which appears not to be previously addressed by researchers.

Consequently, it is understood that these comparisons sustain the initial claim that the AHP can't solve complex problems. At the same time, it supports something that is implicit but not mentioned in this chapter.

In this analysis, it has been discussed and examined the drawbacks and short-coming of AHP, and based on it can be concluded that this method is not prepared to handle scenarios more than those trivial.

It is necessary and fair to point out that AHP is not the only method to be blamed by its inability of modelling and solving complex problems. It also indicates that practically there is not a MCDM method that can cope with this issue.

Future studies should gradually end the promotion of outdated methods and, instead, begin developing innovative MCDM methods (Asadabadi et al. 2018).

For this reason, it is considered that the MCDM activity should be revamped, changing its paradigm. What is needed is to develop an MCDM method to be capable of representing reality accurately, albeit possibly not in 100%, according to our present-day technological knowledge. There is the need to have a sensitivity analysis based on solid grounds, not on preferences as happens nowadays. These authors only hope that this publication can help in making researchers aware of this acute need.

The Need for Improvement of the MCDM Modeling

6

> *Not only does the importance of the criteria determine the importance of the alternatives as in a hierarchy, but also the importance of the alternatives themselves determines the importance of the criteria.*
> Thomas Saaty

Abstract

This chapter proposes a structure for solving MCDM complex problems. It analyses the different relationships between alternatives, their dependency and their modeling.

Keywords

Complex problems · Structure · Dependency · Binary matrix for criteria

6.1 Introduction

Precedent chapters analysed shortcomings and drawbacks of the AHP method. Considering that many new methods have been developed as hybrids (combining two or more methods), for instance, TOPSIS and AHP, it appears that in these cases, a problem is divided into two, both with different purposes. In the pair cited, AHP is used to provide weights for criteria, and TOPSIS is further employed to choose the best alternative.

This procedure transfers some of the inaccuracies and weaknesses of AHP into TOPSIS, since the trade-offs are not suitable for selecting alternatives, and which is worse, for performing the sensitivity analysis, in any MCDM method.

© The Author(s), under exclusive license to Springer Nature Switzerland AG 2021
N. Munier, E. Hontoria, *Uses and Limitations of the AHP Method*, Management for Professionals, https://doi.org/10.1007/978-3-030-60392-2_6

Other than that, most MCDM methods depart from a modeling that ignores many characteristics of scenarios and then solving problems that are poor representation of reality.

In this book, It is proposed a modeling based on the following scheme (see Fig. 6.1).

6.2   Explanation

A problem may consist of a set of definite alternatives and one or more scenarios. As an example, assume a multinational company in the food sector. They are active in grains of several types, as well as milk, meat of different origins and oranges, and own large plots of lands in different countries. Each one of these products constitutes an alternative, and this problem can be solved by different MCDM methods, when considering only one scenario.

However, in this problem intervene other sites or scenarios. That is, in this case, it requests the selection of projects, but also selecting the different sites or scenarios where each project can be developed, in such a way as to get the best project in the most suitable plot.

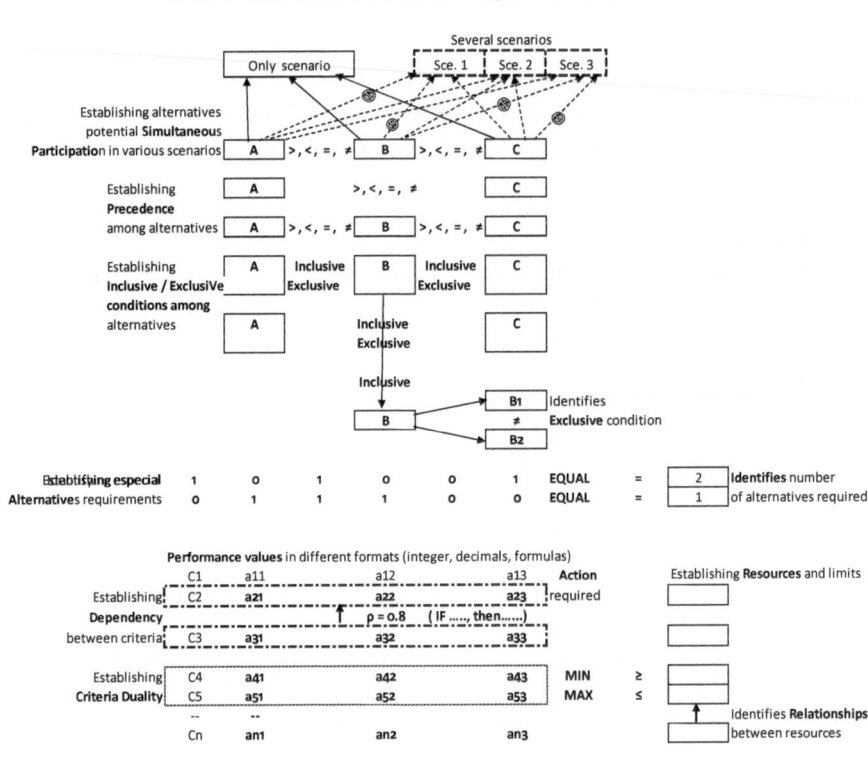

Fig. 6.1 Proposed scheme for modeling

Figure 6.1 displays a framework for modeling a large kind of scenarios, regarding their characteristics, because it considers most of the features found in simple and complex problems. The framework is formed by the typical three main components, alternatives, criteria and performance values, but also includes different mathematical elements that allow to incorporate in the model most characteristics present in the scenarios.

Its main feature is that it works with inequalities and with binary matrices; these two elements are the key for a realist modeling, due to their ability to input diverse occurrences. Therefore, this framework, in addition to the three classical components, incorporates:

- Criteria as a set of inequalities and equalities. This is probably the main reason for the flexibility of this framework. In this way, it is possible to express the convenience to consider a project adequacy to each scenario. For instance, it may show that an alternative A, say wheat, may be in different scenarios simultaneously, because the conditions are good to develop this crop, as happens in the real world.
- Of course, it also shows that the three projects, A, B and C, can be developed in one scenario. But this approach changes when more than one scenario is considered. For example, assume that alternative A, for whatever reasons, cannot be developed in scenario 1 because it is not suitable for this crop; however, it can be grown in scenarios 2 and 3. This can be inputted in the model.
- Alternative B, say meat, cannot be developed in either scenario 1 or 3, but can in 2, while alternative C, say orange grove, can be developed in scenarios 1 and 2, but not in 3.

These are hard restrictions; however, they can be modeled using, for instance, criteria clusters and/or binary matrices, each one corresponding to a scenario and each clustering with their own characteristics, regarding different aspects such as soil quality, rains, pests, export duties, etc.

There could also be conditional restrictions, for instance, that project D that stands for an orange juice concentration plant is dependent on the orange grove. This can be expressed in an inequality such as $C > D$ in one scenario, but also be that $C = D$ in other scenario, meaning that if the orange grove is developed, then the concentration plant has to be also built. These strategies can be modeled using the mathematical symbols $\geq$, $\leq$, $=$ and $\neq$.

- Operators. The operator '$\geq$' can be also used to indicate, for instance, the minimum amount of water needed for each alternative and in each scenario. The '$\leq$' may instruct the software to consider up to a maximum value, for instance, the amount of funds budgeted for each alternative.

In other projects, '$\geq$' may be used to indicate, for instance, the size of the minimum area to be sown of each crop attending economies of scale or for a soil low in nutrients.

The operator '$\leq$' may be used to indicate that the level of air contamination, whatever the project selected, should be less than a certain amount. For instance, in a 1000 MW power plant to indicate, the maximum content of CO_2 should be less than 6 million tons/year.

In another project, the '=' operator can be used to indicate that it is to be used in exact quantity, neither more nor less.

In these examples, it can be appreciated the true significance of using inequalities and how they can be utilized to model many characteristics of a scenario.

Binary matrices, a very important component, allow for operations of dependency between alternatives. It can be established, for instance, when only one project must be selected or when all projects must be considered:

- Dual criteria that permit addressing a very common feature, which is working simultaneously with minimum and maximums on the same criterion and then allowing the method to select any intermediate value
- Resources and their limits
- Conditioning of the 'If…then…'

6.3   Conclusion of This Chapter

After the analysis performed along this book, it is evident that present-day MCDM methods, even when considered mathematically sound, only may model and solve a simplification of real problems. Unfortunately, from this point of view, there has been no progress since there was a switch from the conventional economic models to consider the intervention of other areas. There have been advances in getting better data mainly by using the family of techniques based on fuzzy logic, but nothing has been done to improve modeling, since we are still assuming that problems have hierarchical characteristics, although the introduction of network representations has improved, and that was done decades ago, and even them are unable to model some scenario characteristics.

It is understood that this situation needs to be corrected, and in this section, it is shown that the old practice, based on a hierarchical structure and on equality needs to change. It is believed that using inequalities, as well as abandoning the pair-wise comparison method for eliciting preferences, existent MCDM methods may be modified not in their mathematical structure, but in their modeling, and be able to solve complex problems.

There is also a dire necessity to improve sensitivity analysis which nowadays and from its early days is performed based on dubious and debatable practices; for example, in selecting for variation, only one criterion is chosen because it has the highest weight.

There is a need for furnishing the stakeholders with valuable information regarding the best alternative and ranking, and especially to examine its performance when exogenous factors may affect the best solution, as well as to develop a quantitative and reliable measure of risks, in the later circumstances.

In an effort to collaborate with practitioners in MCDM scenarios, these authors have developed a tool for selecting the most appropriate MCDM method for a particular scenario. This tool, depicted and illustrated in the Appendix, Section A1, considers the present-day state of this discipline.

In Section A2, it is illustrated the two main classes of problems addressed by MCDM, regarding complexity, that is, lineal top-down hierarchy and complex scenarios.

Final Remarks and Conclusion on AHP/ANP Inability for Solving Complex Scenarios

7

Abstract

This chapter synthetizes and tabulates the conclusion of the issues addressed in the precedent six chapters.

Keywords

AHP · ANP · MCDM · Preferences · Researchers · Statistics

7.1 Conclusion from Precedent Chapters

As posted repeatedly, this book was written with the purpose of enumerating and analysing critically the AHP method and demonstrating, albeit not mathematically, that it is unable to solve problems in complex scenarios. The reason is simple: the AHP structure is not prepared to model many aspects present in complex scenarios.

In order to discuss properly about complex problems from the point of view of MCDM, a definition was proposed and, more importantly, explained and illustrated with several examples.

AHP has several weak points, but probably the most notorious is related with its philosophy of ignoring reality in real-world problems and postulating that they can be solved by a DM, working alone or in groups, establishing preferences and assuming that those are enough to model and solve a problem.

The method is grounded on some psychological considerations that may be adequate for the type of scenarios that involve personal, corporate or military problems, but not for scenarios which characteristics cannot be inputted in a vertical structure, and that fundamentally involve people and a myriad of technical, economic, social and environmental problems.

Because of its assumptions, structure and subjectivity, AHP is not flexible; it looks for simplifications and is very much theoretical.

However, it is fair to point out that this method is not the only one to be blamed by their inability of modeling and solving complex problems. The same can be said for most MCDM methods, because they are not prepared to consider advanced characteristics in some scenarios, and future studies should gradually end the promotion of outdated methods and instead begin developing innovative MCDM methods (Asadabadi et al. 2018).

This research into AHP is based on a compressive bibliographic consultation involving more than 130 authors and has detected its main problems, which in order of importance as per these authors are:

1. The hierarchical structure, which is not flexible and does not permit modeling complex scenarios
2. The pair-wise comparison system and further evaluation of criteria dominance and the impossibility to justify decisions
3. The assumption that preferences are constant
4. The assumption that a project—a system—can be disaggregated
5. The obligation to look for transitivity, even with an allowance of 10%

7.2   What Needs to Be Done?

After the analysis performed in this book, it is evident that present-day MCDM methods, even when considered mathematically sound, are capable of modeling and solving real-world problems only in a simplified way and by ignoring certain inherent characteristics. Unfortunately, from this point of view, there has been no too much progress since there was a switch from the conventional economic models to consider the intervention of other areas. There have been advances in getting better data mainly by using the family of techniques based on fuzzy logic, but nothing has been done to improve modeling.

It is understood that this situation needs to be corrected and that the old practice, based on a hierarchical structure, needs to change. It is believed that using inequalities, as well as abandoning the pair-wise comparison method for eliciting preferences, existent MCDM methods may be modified, not in their mathematical structure, but in their modeling, and be able to solve complex problems.

There is also a dire necessity to improve sensitivity analysis which nowadays and from its early days is performed based on dubious and debatable practices; for example, in selecting for variation, only one criterion is chosen because it has the highest weight.

There is a necessity for furnishing the stakeholders with valuable information regarding the best alternative and ranking, and especially to examine its performance when exogenous factors may affect the best solution, as well as to develop a quantitative and reliable measure of risks, in the later circumstances.

7.3   Claims Against the AHP Model Versus Researchers' Opinions

The content of this book aims at a comparison between these authors research and examination of AHP and those from about 105 researchers, scholars, professors and practitioners from around the world, since the method was published in the 1970s. The fact that after more than 40 years the original doubts and criticism subsist and more doubts and claims have been published indicates that AHP defenders were not able to put these critiques to rest. Another revealing reality is that the amount of criticism on this method overshadows the criticism aimed to other MCDM methods such as SAW, PROMETHEE, ELECTRE, TOPSIS, etc.

Naturally, in examining hundreds of papers published by people using AHP, of course, most of them assert a lot of good attributes of the method but only in words, since nobody demonstrate them. For instance, it is very common to read that AHP can be applied to complex undertakings, which as shown in this book is not true, or that the method can be validated, which is absurd since nobody knows which the true result of a scenario is, or praising the advantage of the method in disaggregating a project and solving each part separately, procedure that does not resist a minimum analysis, because it violates the system theory.

Naturally, there have been people like Saaty himself and his friends and collaborators Vargas and Harker who have produced serious technical papers based on mathematical considerations to rebut the criticism. However, none of them contradicts the findings of other researchers, and aspects such as rank reversal remains unexplained and unsolved, since 1983.

A characteristic of Saaty and his backers is that they use an obscure and unclear language and introduce issues without a formal explanation. One of them is the concept of feedback. These authors have consulted about it to three very well-known experts in AHP and got no rational and convincing answers.

7.4   A Comparison Table and Statistics

To facilitate the comparison of the agreements or disagreements between these authors' opinions and those from other colleagues, each subject is followed by a Table 7.1, with contrasts vis-à-vis points of views.

Researchers addressing a particular subject are identified by name (only the first author in each case), which indicates the interest on each one. The last column indicates the number of researchers that expresses their opinion, in pro and in con on each subject; this gives an idea of the interest of researchers for each one and from there its importance. Those underlined names are the method defenders.

Table 7.1 Researchers commenting on each subject

Subjects	Researchers with opinions on each subject	Number of researchers addressing each subject
5.1 Introduction—structure of this chapter		
5.1.1 General criticism of these method	*Dyer, Harker* et al., *Ishizaka, Smith* et al., *Rodrigues* et al., *Bana e Costa* et al., *Asadabadi* et al., *Kahneman., Salvia* et al., *Jimenez* et al., *Perez* et al., *Zardari* et al., *Saaty, Ghazinoory.*, *Sun* et al.	15
5.2.1 Pair-wise method and its application in AHP	*Goepel, Bozóki* et al.	2
5.2.2 The pair-wise method in AHP constructs artificial relationships	*Kunsch, Tomashevskii, Song* et al., *Schoner*	4
5.2.3 Criteria preferences must consider alternatives	*Salo* et al., *Hulkower* et al., *Dyer, Russo* et al.	4
5.2.4. The ambiguity of pair-wise comparisons in AHP	*Köksalan* et al., *Salo* et al., *Kunsch, Hulkower* et al., *Barakos, Hazelrigg, Sehra* et al., *Saaty*	8
5.2.5 Modeling scenarios	*Ksenija* et al., *Velasquez* et al., *Carlsson* et al., *MIT*	4
5.2.6 Report to stakeholders and lack of rational answers from DMs	*DTLR, Habenicht* et al., *Koen*	3
5.2.7 AHP incapacity to solve complex problems	*Satell*	1
5.2.8 Criteria independency	*Barba-Romero, Kasperczyk* et al., *Improta* et al., *Saaty, Lee, Öztürk, Yüksel* et al.	7
5.2.9 Quantifying preferences	*Ishizaka* et al., *Saaty, Bernasconi* et al., *Buchanan* et al., *Barzilai, Dodd* et al., *French, Song* et al.	7
5.2.10 Quantitative data		
5.2.11 The DM needs to be an expert in different fields	*Velasquez* et al., *Konidari* et al.	2
5.2.12 The AHP is forced to correct his own decisions due to a formula	*Cole*	1
5.2.13 It may be wrong to consider that relative importance between criteria is constant	*Schokkaert*	1
5.2.14 Which are the limits for the AHP's fundamental scale?	*Asadabadi* et al.	1

(continued)

Table 7.1 (continued)

Subjects	Researchers with opinions on each subject	Number of researchers addressing each subject
5.2.15 A preference and the practical meaning of its inverse value		
5.2.16 Determining relative criteria trade-offs is wrong if alternatives are not considered	*Perez* et al.	1
5.2.17 The DM preferences do not consider the real world; they only exist in his own universe	*McCaughey* et al., *Danielson* et al.	21
5.2.18 The need for the DM to correct himself with 'small' changes	*Fishburn, Asadabadi* et al.	2
5.2.19 Normalization of priority values	*Ishizaka, Checkland*	2
5.2.20 The selection of the eigenvector method	*Bana e Costa, Ishizaka* et al., *Farkas*	3
5.2.21 Criteria trade-offs are assumed to be weights	*Stanujkic* et al., *Grupp* et al., *Triantaphyllou, Vujičić* et al., *Kazibudzki, Scboner* et al., *DTLR, Cinelli* et al., *Zanakis* et al., *Yeh, Moshkovich* et al., *Zardari* et al., *Belton* et al., *Ma* et al., *Farkas*	15
5.2.22 A projects is a system; it can't be disaggregated	*DTLR, Ishizaka* et al., *Hobbs* et al., *Uni>ersia, Hullkower* et al.	5
5.2.23 Sensitivity analysis and the way it is performed using AHP	*Triantaphyllou*	1
5.2.24 Rank reversal (RR) in AHP	*Pérez, Pérez* et al., *Salo* et al., *Forman, Farkas*	5
5.2.25 AHP can't consider time dependency in a portfolio of projects	*Saaty, Lin* et al., *Hashekhani*	3
5.2.26 AHP can't consider the effects of future events (Limiters)	*Hashemkhani* et al.	1
5.2.27 A hierarchy structure is not suitable to model real-world problems	*Öztürk, Song* et al., *Cillers*	3
5.2.28 One criterion may place a restriction on the optimal achievement of another criterion		

(continued)

Table 7.1 (continued)

Subjects	Researchers with opinions on each subject	Number of researchers addressing each subject
5.2.29 Problem structuring must be flexible enough to incorporate most scenario characteristics	*Belton* et al.	1
5.2.30 In AHP, the assumption that a criterion impacts and modifies another criterion is erroneous	*Liu* et al., *Perez* et al.	2

Total researchers: 124

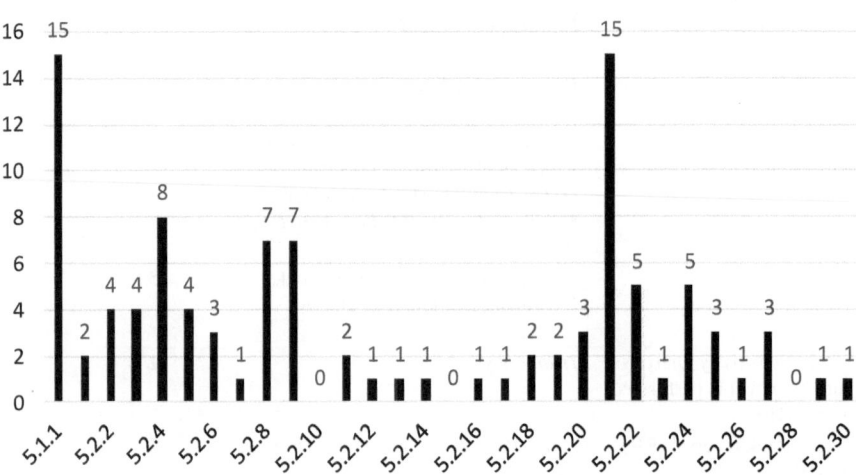

Fig. 7.1 Bar chart showing researcher's interest on each subject

Notice that the subjects with the largest proportion of opinions and comments are 'General criticism on AHP' and 'Criteria trade-offs' (which AHP assumes to be equivalent to criteria weights), followed by 'The ambiguity of pair-wise comparisons', 'Criteria independency' and 'Quantifying preferences'. Figure 7.1 graphically reflects these results.

The statistics show that out of the 30 subjects raised by these authors, 90 percent were previously addressed by researchers, and 10 percent are new and originated by these authors.

Out of 105 researchers, 101 are supporting these authors and 4 not what they postulate, the latter exclusively coming from three researchers, Thomas Saaty, Luis Vargas and Patrick Harker, who happens to be Saaty's co-authors and collaborators.

That is, out of the author of the method and his two colleagues, no researcher comes to justify it.

The authors of this book tested for responses to their questions while working in a scientific forum where AHP method was amply addressed, and out of 56 readers of their questions, no defender came to justify the method.

In this book, taken into account the number of researchers that address each subject, the largest number of opinions refer to two main subjects 'General criticism' (5.1.1), with 15 negative opinions and end 2 positives. The other most visited and discussed subject are 'Criteria trade-offs' *(5.2.21)* (15 negative opinions), 'Pairwise comparison' (7 negative and 1 positive), 'Criteria independency' *(5.2.8)* (6 negative and 1 positive) and 'Quantification of preferences' *(5.2.9)* (6 negatives and 1 positive). As can be seen, there is in total a very large difference: 95.2% negative and 4.8% positives. These percentages speak by themselves.

As a conclusion, and considering the overwhelming negative opinions versus the positive ones, with prestigious and very well-known researchers and scholars from both sides in the MCDM field, these authors conclude that in addition to the many shortcomings of this method, there is strong evidence of the inability of this method to solve complex scenarios.

Appendix

A1. Tool for Selecting the Most Appropriate MCDM Method to Solve a Problem

It is a matrix where the prospective MCDM methods are in columns, and 54 different potential characteristics of a scenario in rows (Table A1). This matrix is in binary format, and then, a '1' identifies when a requirement is met by a method, and '0' when it does not.

The methods are arranged in a capacity order, from left, being the method with the least capacity, to the right being the method with the largest capacity. This matrix is called the *Membership matrix* and it is used as a reference; consequently, it does not change, unless new methods and/or new requirements are added or when one method is improved by new features.

There is another matrix which is called the *Working matrix* (Table A2) that initially is a replication of the membership matrix.

Its use is simple since the DM has only to perform two actions:
1. He has to go to the column labelled *Scenario characteristics* and starting with the first one, determines if his project or scenario has this requirement.

   For instance, the first one is *Simple scenario*, and if the project meets this classification, then he puts a (1), which will show in red, in the column *Your problem and needs.*

   He continues downwards until finding another requirement that the project must meet, for instance, in row 20 *Independent criteria,* and then places a (1).

   The DM continues until the next one which could be for instance *Clustering*, in row 28, and then places another (1).

   Next one is, say, *Necessity to evaluate criteria relative importance,* in row 46, where the DM places a (1), and then he finds that there are no more memberships until the end.
2. The DM delets all (1 s), that are in rows where there is not a membership, that is, in this case he deletes 50 (1 s), *just the (1 s), not the rows.* That is all.

Looking at the bottom of the list the DM realizes that he has four rows in red. He needs only to look at the third row labelled 'Most appropriate method, the lower the better'.

© The Author(s), under exclusive license to Springer Nature Switzerland AG 2021 105
N. Munier, E. Hontoria, *Uses and Limitations of the AHP Method*, Management for Professionals, https://doi.org/10.1007/978-3-030-60392-2

Table A1   Most frequent characteristics that appear in scenarios

Characteristic ID	Characteristic
1	Simple scenario
2	Several scenarios
3	An alternative may simultaneously be in different scenarios
4	Simple objective
5	Many objectives
6	No rank reversal
7	Necessity to have an optimal solution
8	Several DMs (Group Decision-Making)
9	Easiness to change the initial matrix values
10	Large project involving people consultation
11	Linguistic initial matrix
12	Qualitative criteria
13	Quantitative criteria
14	Using a particular normalization procedure
15	Using any normalization procedure
16	Independent alternatives
17	Relationships between alternatives
18	Dependency between alternatives
19	Many criteria
20	Independent criteria (Compensatory methods)
21	Relationships among criteria
22	Necessity of knowing criteria validity range
23	Correlation between criteria
24	Necessity to express criteria positive actions (benefits)
25	Necessity to express criteria negative actions (costs)
26	Criteria duality
27	Reasonable preparation time and computing time
28	Clustering
29	Necessity to consider externalities
30	Necessity to consider joint ventures
31	Necessity to consider use of resources
32	Necessity to use thresholds in resources
33	Necessity to link resources
34	Performance values as linear functions
35	Performance values as non-linear functions
36	Integer performance values
37	Decimal performance values
38	Objective performance values
39	Subjective performance values
40	Performance values expressed in mathematical formulas
41	Performance values in binary format
42	Negative performance values
43	Result needed in integers
44	Result needed in decimals
45	Result needed in binary format
46	Necessity to evaluate criteria relative importance
47	Want to use subjective weights
48	Want to use objective weights
49	All criteria with the same weight
50	Sensitivity analysis (SA) with weights
51	Sensitivity analysis (SA) with marginal values
52	Sensitivity analysis (SA) considering simultaneously all criteria
53	Necessity to have graphics in sensitivity analysis (SA)
54	Not theoretical complexity

Table A2 Membership matrix, matching criteria and MCDM methods

SAW	AHP	TOP	VIK	PRO	MOO	ELE	ANP	LPr	SIM	
1	1	1	1	1	1	1	1	1	1	10
	1								1	2
									1	1
								1		1
1	1	1	1	1	1	1	1		1	9
						1	1	1	1	3
								1		1
	1				1*				1	2
1		1	1	1	1	1		1	1	8
1		1	1	1	1	1		1	1	8
	1						1			2
1	1	1	1	1	1	1	1	1	1	10
		1	1	1	1	1		1	1	2
	1	1			1		1			4
1								1	1	3
	1						1			2
								1	1	1
								1	1	1
1		1	1	1	1	1		1	1	8
1	1	1					1			4
				1				1	1	3
		1	1		1		1	1	1	7
								1	1	1
1		1		1		1		1	1	6
		1		1				1	1	4
1		1	1	1	1	1		1	1	8
1		1		1				1	1	5
	1						1	1	1	4
1	1	1	1	1	1	1	1	1	1	10
								1	1	2
				1				1	1	3
								1	1	1
								1	1	1
								1	1	1
								1	1	1
1		1	1	1	1	1		1	1	8
1	1	1	1	1	1	1	1	1	1	10
1		1	1	1	1	1		1	1	8
1	1	1	1	1	1	1	1	1	1	10
									1	1
									1	1
									1	1
1	1	1	1	1	1	1	1	1	1	10
									1	1
1	1	1	1	1	1	1	1		1	9
1	1	1	1	1	1	1	1			8
1		1	1	1	1	1		1	1	8
1		1	1	1	1	1				6
1	1	1	1	1	1	1	1			8
									1	1
									1	1
				1					1	2
1	1	1		1						4
Requirement										
Match req. 23	18	25	19	26	20	21	17	26	44	
Best										

Consequently, the DM looks for the lowest and then determines to what method it belongs to.

The DM can realize that according to the characteristics of the problem, the most appropriate method is AHP with a score of (0), that is, fully complying with the requirements, SAW with (1) (less compliance), and ANP, also with a score of (1) (less compliance).

The red arrows illustrate this procedure.

This is not an exact science; this tool was built based on personal experience after examining hundreds of projects published in Internet and in scientific journals.

Table A3 Working matrix to find the most appropriate MCDM method for a given scenario

		SAW	AHP	TOP	VIK	PRO	MOO	ELE	ANP	LPr	SIM	
1	1	1	1	1	1	1	1	1	1	1	1	10
2												
3												
4												
5												
6												
7												
8												
9												
10												
11												
12												
13												
14												
15												
16												
17												
18												
19												
20	1	1	1	1								3
21												
22												
23												
24												
25												
26												
27												
28	1		1						1	1	1	4
29												
30												
31												
32												
33												
34												
35												
36												
37												
38												
39												
40												
41												
42												
43												
44												
45												
46	1	1	1	1	1	1	1	1	1			8
47												
48												
49												
50												
51												
52												
53												
54												
Requirement	4											
Match req.	4	3	4	3	2	2	2	2	3	2	2	
Best		1	0	1	2	2	2	2	1	2	2	

A2. The Two Main Classes of Projects Where MCDM Is Applied

Direct Projects – Solved by Preferences – Characteristics

Normally hierarchical structured scenarios, lineal top-down personal decisions, independent subjective criteria and sub-criteria, clustering involve a single scenario.

Structure: Lineal Top-Down Hierarchy

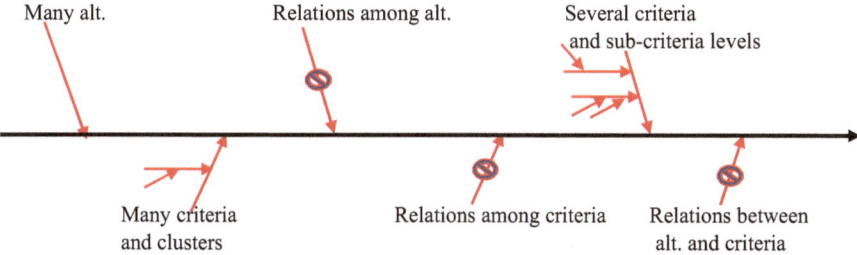

Complex Projects – Characteristics

Normally network structured scenarios, direct and indirect relationships, large quantities of independent and dependent alternatives and criteria, inclusive and exclusive relationships, predominance of objective data, mixed with subjective estimates, large quantity of people affected, time participation, time-dependent alternatives, strong constrained criteria, suppositional questions (What…, if…,), conditional features (If…, then…), probabilistic values, clustering, involve several scenarios, special conditions between alternatives and criteria, result-oriented analysis, performance measurement.

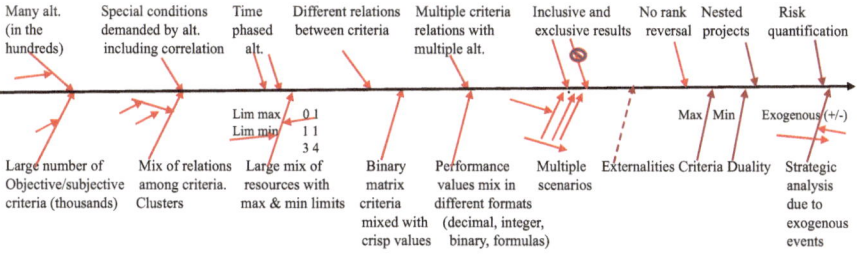

Glossary and Explained Definition of Terms Mentioned in this Book[1]

Analytical Hierarchy Process (AHP) (Saaty 1994)   A MCDM method based in decomposing a project according to a hierarchy.

Analytical Network Process (ANP) (Saaty 2008b)   A MCDM method based on networking, and then, able to represent multiple relationships

Aggregation   Summation of partial weights.

Alternatives   The projects, options, indicators and so on to attain the main objective or goal.

Attributes in criteria   In MCDM, these are the characteristics of the criteria. Since criteria are formed by numbers, or performance values, attributes normally refer to them. Attributes may include a number of performance values, type, for instance integers, decimals or binary, values for all alternatives and so on. The most important attribute is the amount or quantity of information that each criterion possesses. This is given by the dispersion of the values, and this is fundamental because it is the only way a criterion can evaluate alternatives. If a criterion has close values, for instance 17, 19, 23, 20, then the dispersion is low and the amount of information is poor. Another criterion having, for instance, values 15, 28, 5, 34, because the dispersion of its values provides a larger amount of information. This is easy to understand considering a die; if it is perfectly balanced, then all six numbers have the same probability to appear when the die is cast, consequently, for a player it gives no information.Quantity of information is measured by the criterion entropy.

Axiom   It is a statement which is so evident that it is considered to be true, with no need for demonstrations, and that can be used for further reasoning. Probably, the most known axiom in MCDM is transitivity, where if A > B, and B > C, then A > C. Transitivity among the DM judgements is the core of the AHP method.

Benefit/Cost ratio   A procedure that compares benefits from projects and funds used to developing it. Since these costs and benefits take place at different time

---

[1]By way of a glossary, the following text contains the explained author's definitions of terms related to MCDM. They are explained in plain English, that is, not using strict mathematical language, and then, with a simplified exact meaning and scope. Where necessary for better comprehension, they are complemented with examples.

N. Munier, E. Hontoria, *Uses and Limitations of the AHP Method*, Management for Professionals, https://doi.org/10.1007/978-3-030-60392-2

periods, all values need to be brought to present values which are done using discounted rates.

Binary matrix   A matrix formed with 1 s and 0 s.

Carbon emission bonds   It is an international mechanism by which a company in which CO_2 emissions have been reduced or that has not reached its allowable contamination levels of that gas is allowed to offer for sale his credit to another company, anywhere in the planet.

Choice (in MCDM)   Selection of the best alternative.

Clusters   Set of alternatives or criteria. For instance, in an indicators' selection problem to measure country development, indicators are grouped in categories, such as Economics, Environment, Social issues and so on. Within each category, there could be several indicators that form a cluster (see table). It is also possible to have clusters of criteria. For instance, a criterion for environmental contamination may include these sub-criteria: CO_2, NO_x, Particulate, Biochemical Oxygen Demand (BOD) and so on.

Example of categories and cluster of alternatives								
Categories								
Economy			*Environment*			*Social issues*		
Indicators in clusters								
GDP per capita (US$/ capita)	Disposable income (US$)	Estimated oil reserve (billions of barrels)	Forested area (km²)	Erosion (km²)	CO_2 contamination (tons)	Water consumption per capita (litre/ day-person)	Gini index (%)	Crime (%)
Criteria								

Compensation   When the variation of a criterion weight affects the weights of other criteria.For instance, in AHP, all priorities add 1. Therefore, the variation of the weight in one criterion affects all of the others.

Consistency Index (CI)   In AHP is the average sum of errors in the values obtained by the eigenvector method. For perfect consistency CI = 0 and for accepted inconsistency $\dfrac{CI}{RI} < 0.1$ .This index is computed as follows: $CI = \dfrac{(\lambda \max - n)}{n-1}$.

Consistency Ratio (CR)   In AHP, it is the ratio between CI and RI (Random Index) which is the random index as an average of CI from 500 matrices of each size and obtained by simulation.For instance, if for n = 10, RI is 1.49 (from a table); this value indicates the average of consistency regarding the matrix n = 10. If for instance CI =0.2, then: $CR = \dfrac{CI}{RI} = \dfrac{0.2}{1.49} = 0.13$ According to Tomashevskii (2015), the CR is not an acceptable (EM) error indicator.

Consistent matrix

If a matrix is perfectly consistent, then Eq. (1) holds:

$$a_{ij} \cdot a_{jk} = a_{ik}$$

Consider the following matrix:

Criteria	C1	C2	C3
C1	a_{11}	a_{12}	a_{13}
C2	a_{21}	a_{22}	a_{23}
C3	a_{31}	a_{32}	a_{33}

Suppose that the preferences of the DM for criteria are as follows:
C1 = 3 C2,
C1 = 6 C3,
C2 = 2 C3.
See in table below the relationship of these statements:
We have to prove that formula (1) holds, then :

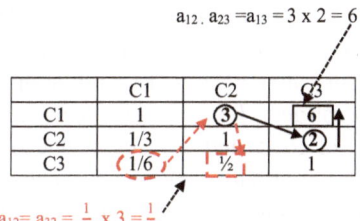

$$a_{12} \cdot a_{23} = a_{13} = 3 \times 2 = 6$$

	C1	C2	C3
C1	1	3	6
C2	1/3	1	2
C3	1/6	½	1

Also: $$a_{31} \cdot a_{12} = a_{32} = \frac{1}{6} \times 3 = \frac{1}{2}$$

Consequently, this matrix is consistent.

Correlation   Mathematical operation that indicates the extent to which two variables vary at the same time.

Criteria duality   A criterion that can be subject to two opposite actions, such as maximization and minimization. Both criteria are identical, that is, both have the same performance values. Differ by the corresponding mathematical operator, defined by the action, and the respective limits.

Criterion   Condition that alternatives must comply. Each criterion is formed by a series of values (performance values), each one corresponding to an alternative.

Crisp value   A performance value obtained through a fuzzy operation.

Decision-Maker (DM)   The person in charge of an MCDM analysis. The most important component of the system.

Dependency   Between alternatives, when some of them constraint another.

Design   'Fabricate' new alternatives as a function of criteria.

Disposable income   It is the income someone perceives after deducting taxes and other fiscal obligations, and then, it is the net amount to be spent on whatever goods and services.

Equality   A mathematical statement indicating equal values of two expressions. The result is only one. Its representation is a line.

ELECTRE method (Roy 1968)   An MCDM method based on outranking.

Eigenvector method (EM)   (Very elemental definition!). It is a vector of a square matrix. An eigenvector is a vector that does not change direction when it is stretched when mapped by a matrix. It has a non-zero eigenvalue that expresses the magnitude of the change. The eigenvector method could not deliver results that are Pareto optimal.

Entropy (S)   Measures quantity of information contained in each criterion. The highest the entropy, the lower its information content. Thus, quantity of information: $1 - S$, with entropy values normalized.

Evaluation   In MCDM, it involves determining the relative importance of the alternatives to comply with all criteria. The evaluation produces the scores.

Externalities   Aspects that do not have a market value, for instance noise, erosion, natural resources depletion

Fuzzy process (Zadeh 1965)   Boolean logic establishes a binary value for something for instance (1) for clean and (0) for dirty. Fuzzy logic approaches the issue differently by establishing 'degrees of cleanliness'

Geometric average   In a problem with 'n' alternatives, it refers to the criteria. The geometric value of each criterion is the n-root of the product of its performance values. Then, for a criterion with values: 25, 39, 56, 19 its geometric average is: $\sqrt[4]{25 \times 39 \times 56 \times 19} = 31.91$. The geometric mean is Pareto optimum.

Gini Index   It is a measure of inequality of wealth in the inhabitants of a country.

GDP: Gross Domestic Product   It is a worldwide used measure to gauge in monetary units the number of goods and services produced in a country in a year.

Group decision-making   A group of decision-makers, stakeholders and interested parties aiming at reaching an agreement regarding a MCDM problem, for instance, to decide about criteria and their number, or to consolidate preferences on pair-wise comparisons. Diverse MCDM methods allow for group decision-making using different procedures.

Heuristic method   In MCDM, it is a mathematical method very often employed to solve a problem through reasoning and experience, to find a compromise solution. It is employed when the optimal solution involves complex mathematical procedures or does not exist. Almost most MCDM methods are heuristic, while Linear Programming mono – objective gives optimal solutions. SIMUS, a MCDM method is heuristic but is mainly based in Linear Programming.

Hierarchy   In AHP, the decomposition of a project in goal, criteria and alternatives. The first at the top is the goal or objective. It is a top-down approach similar to the lineal organization and the military chain of command

Impact   In MCDM is the influence of one element over another. For instance, heat normally impacts certain food, and this is reason for them to be refrigerated.

Independent   In criteria and alternatives, those which are not related. For instance, *Available water* and *Manpower*, as opposite to related, for instance, *Production cost* and *Rejects during production.*

Indifferent criterion   It is a criterion with identical or very similar performance values. It is called 'indifferent' because it can be eliminated since its contribution to alternatives selection is null. In this case, its entropy is close to 1, and the corresponding quantity of information is close to 0.

Inequality   It is a mathematical statement that indicates that one expression or quantity is *larger than* or *lesser than* another. There could be several answers because it includes a space. In MCDM methods, they are the base for Linear Programming.

Initial Decision Matrix   The matrix (square in AHP), but not necessarily in other methods, formed by alternatives, criteria, performance values and limits to criteria.

Internal Rate of Return (IRR)   It is a measure of a return produced or assumed by an investment. Together with the Net Present value, they are the main financial indicators. The IRR is the discount rate at which the NPV turns from positive to negative, that is, the intersection of the NPV curve with the abscissa that indicates the discount rates.

Interrelationships   In MCDM, it is common to find that criteria are connected or related among them. The same may happen between alternatives. For instance, a criterion *Weather in a working site* may be connected with *Work performed in a certain period of time.* In AHP, for instance, criteria are connected with alternatives, and in ANP, everything is connected to everything. However, relationship does not necessarily indicate effect or impact from element A to element B.

Joint venture   Normally seen in tenders for bids when two or more companies, normally independent and producing different products, make a joint presentation, usually one complementing the other. For instance, in large projects, the supply of large electrical transformers is joined by a company fabricating electric control boards, or a steam turbine manufacturer with a large constructor of electric generators. This system normally has advantages for the entity making the call, such as lower prices, but mainly because there will be a joint responsibility for the good performance of such complicate and delicate equipment when working together, one depending on the other.

Lambda (λ)   In the eigenvector mathematical model, this is the eigenvalue, which is the larger value for the right eigenvector. In AHP, it indicates the difference between the ideal λ which value is equal to the order of the matrix and the value obtained by the eigenvector calculation.

Limiters   Refers to actions that can take place in the future and affect the decision-making. For instance, the influence that exogenous parameters may have in the best selection. An example could be the necessity to determine how the variation in the exchange rate may have in the export of our best product. This factor was not considered in the initial decision matrix because it is exogenous and on which the company does not have any control, but it can have a large incidence and may be to put at risk the performance of our best selection.

Linear Programming (LP) (Kantorovich 1939; Dantzig 1948)   It is a mathematical technique designed to optimize (maximize or minimize) a linear output (objective), subject to many linear constraints (criteria) expressed as inequalities and considering the best allocation of resources, articulated as independent terms of inequalities. Every problem treated using LP delivers two solutions, from resolving the initial matrix and its inverse: the Primal problem that gives the scores of each alternative, which is the main purpose of an MCDM scenario, and the Dual problem that provides the marginal values or shadow prices of each significant criteria, which is fundamental to perform a sensitivity analysis. In both cases, it computes the value of the objective, which coincides in both problems. An Excel add-in, named *Solver*, can solve very rapidly even large PL problems and delivers the two solutions at the same time. Solver uses a software called *Simplex* that finds the solutions.

Marginal contribution or shadow prices   They provide the marginal contribution of criteria, which shows how much the objective function increases/decreases by a unit increment/decrement of the corresponding resource. According to some researchers, a very interesting property of shadow prices is that they can quantify externalities. For a certain significant criterion, its shadow price remains constant, as long as the increases or decreases remain within a certain range that is also delivered by LP.

Mathematical symbols   Used in Linear Programming (LP), such as: '$\geq$' (greater or equal than), '$\leq$' (smaller or equal than), '$=$' (equal than). They are used in LP because this MCDM method works with inequalities. The result is normally expressed using the '$\succeq$' (preferred to) operator, used to indicate the relative importance in a ranking.

Model   A mathematical representation that is a simplification of reality. It must reflect reality as close as possible including the special characteristics, demands and conditions of a scenario.

Modelling a scenario   Building an initial decision matrix reproducing as close as possible a scenario characteristic. It is starting point for MCDM.

Multi Criteria Decision Making (MCDM)   Given a certain number of alternatives, which are subject to the compliance of several conditions or criteria, it consists in finding the alternatives that best comply with them all. There could be many solutions to this problem, and the ideal would be to find the optimal one, that is, a solution that can't be improved, or that it is better than any other. This is normally impossible because the criteria may have contradicting issues such as maximize the benefit and at the same time minimize the cost. For this reason, all the dozens of MCDM methods are heuristic and try to find a compromise solution instead of the optimal one, that is, a solution that satisfies the decision-makers.

Natural Capital   Is the remaining stock of natural resources of a country, such as water, forests, minerals and so on.

Net Present Value (NPV)   It is the present worth of cash flows occurring at different times. That is, inflows and outflows in future years and brought to the present, considering that the further the period considered for a sum of money, the less its present value.

Normalization   Mathematical operation to allow adding up criteria with different units. It is an essential mechanism and allowing to compare heterogeneous data.

Ordering   Assume a series of projects identified by numbers sorted in increasing sequence, like this: **1**, 2, 3, 4. Suppose that the rank obtained by a MCDM method is: 3, **1**, 2, 4. That is, the most important is project 3, following by projects 1, 2 and 4. The order is the position of the sort value respect to rank. Thus, project 1 in sorting is in the <u>second</u> place in the rank; number 2 in sorting is in the <u>third</u>

place in the rank; number 3 in sorting is in the <u>first</u> position in the rank and so on. Consequently, the order is 2, 3, 1, 5.

Objective weights   Weights for criteria obtained using the initial data and without human intervention, such as entropy.

Objective   Main aim of the project. Objectives may be indefinite, such as *Improve quality of life,* or definite, such as *Produce at least 50 units in total.*

Outranking   A comparison between alternative performance values and computing their differences.

Pair-wise comparison (Thurstone 1927)   The comparison of two criteria or two alternatives, with respect to something common to both.

Portfolio of projects   In MCDM, it is a collection of projects, plans, options, alternatives to be executed in a certain period of time. This collection not necessarily has to be homogenous, since it may involve infrastructure projects, social plans, political strategies, sustainable issues and so on.

Pareto efficiency or Pareto optimal   Very broadly: in a series of results which are considered optimal, a value cannot be improved without worsening another.

Pay-back period   Length of time necessary to recover capital invested in a project.

Precedence   Priority of one thing over another. In MCDM, it is interpreted as the necessity of executing a task before performing another task. This is found in networks where one node is connected to another through a one sense, or direct arrow. As an example, there is precedence when an alternative such as *execution of rain-storm sewerage in a street* must precede *changing electric wiring for lighting.* Both are independent, but the first must precede the second.

Performance values   Values that conform to the initial decision matrix. They can be integers, binary, decimals or formulas.

Preferences   When a DM, comparing two criteria or two alternatives, chooses one over the other. However, the DM also exerts his/her preferences when examining the final result from the MCDM process. He may prefer to choose the second-best alternative because at his judgement the best alternative is too sensible to potential changes in some parameters. In AHP, preferences are based on the DM intuition.

Priorities vector   The vector obtained from the eigenvector method, in the AHP method.

Project   Same as alternatives or options.

PROMETHEE (Brans and Vincke 1985)   MCDM method that works comparing a pair of alternatives with respect to a criterion. Uses thresholds and transference functions.

Random Index (RI)   In AHP, it is an index developed by Saaty obtained by simulation and used to determine matrix consistency. Generated by running 500 times different matrices of different sizes and using a matrix with random values for each size. Alonso et al. (2005) appropriately define it as *the average value if CI for random matrices using the Saaty table.*

Ranking   Order of importance of the alternatives corresponding to the value of their scores. In MCDM, the ranking is computed from the best alternative that with the highest score, and in decreasing order to the alternative with the lowest score. This convention applies for both maximizing and minimizing an objective.

Rank Reversal (RR)   In an MCDM result, it is the modification of a ranking when adding a new alternative or by deletion of an existing one, and producing unexpected changes in the ranking, when in reality there is no reason. For instance, in a ranking C > B > A, adding alternative D may produce this ranking: A > C > B > D. Notice that in the new ranking, A, which was last in the original ranking, in now the first, and observe that the order of the ranking has changed for those irrelevant or dominated alternatives, since now A > C, when in the original C > A.However, if the ranking changes something as this: C > D > B > A, then there is no RR since the addition of D respected the independence of irrelevant alternatives. See that the original order of C > B > A holds.

Reality   According to the dictionary, it is the world or the state of things as they actually exist, as opposed to an idealistic or notional idea of them. This is one of the most important concepts in MCDM since it works with scenarios, projects, undertakings that are real, even when they are not based in economic goal or in physical matters, for instance, a portfolio of social projects to fight drugs, to decrease crime, to combat contamination. All of these are tangible issues not because of the subject but of its effects on people, and these effects can be identified and appraised.

Resources   The labour, funds, equipment and materials that are necessary to develop a project. Since resources are not infinite, whatever the project, they put a restriction to criteria. These restrictions may be also expressed as limits, such as those used in environment, in economics or in any other field and are considered in the same conditions as resources, because they constrain the scope of the criteria. For these reasons, resources are normally expressed through quantities, and then, we have euros or dollars for a criterion related to a budget, or expressed for instance in parts per million (ppm) for a criterion related to particulate contamination, or number of deaths for a criterion related to new-borns survival and so on. Since criteria may show a variety of units, this is the reason by which normalization is needed.

Restriction   In MCDM, it is a limit or a threshold that establishes the scope of the criterion.

Risk   It is the product of the probability of occurrence of an event and the impact produced by it. As an example, the risk of an earthquake in an area is the product of the probability of its occurrence and the impact it can produce on people, buildings, constructions, animals and so on.Probability is generally derived from statistics and damage appraised for different structures and not only measured in monetary units. For instance, in the accident that a tsunami provoked by flooding the Fukushima nuclear power plant in Japan, or the accident for the explosion of a reactor in Chernobyl in Russia, the risk most possible was related to the effect of radiations from the damaged installations.Earthquakes intensities are measured by the Richter scale, and then, the impact, and the corresponding risk, is proportional to the intensity expressed by this scale. This is an aspect always considered in high-rises construction in earthquake-prone areas, as for example, Japan and Mexico and Turkey.

Robustness   The persistence of the ranking when there are variations in the criteria. This is related to marginal contribution of criteria.

Saaty's Fundamental Scale   An absolute scale used to quantify preferences.

SAW: Simple Additive Weighting (Churchman et al. 1957)   It is the sum of the products of performance values by the criterion respective weight. It was one of the first MCDM methods and it is still used because of its easiness, although its reliability is doubtful.

Scenario   It is the general name for a particular frame in decision-making, involving the objective, the set of alternatives, the set of criteria, resources and attributable characteristics that constitute the object under study.

Score   Value that each alternative takes and the end of the process. When these scores are ordered according to their importance (the higher the better), they constitute a ranking.

Sensitivity analysis   Mathematical process that analyses changes in the output due to variations in the input. In MCDM, sensitivity analysis examines how a ranking (output) keeps its order when some specific criteria (input) values change. For instance, we have this ranking D > A > C > B.Assume for instance that the best alternative D is dependent on criterion *Exchange rate US Dollar/ Euro*. Sensitivity analysis examines how variations of the exchange rate between these two currencies may affect the ranking, for instance causing alternative D to resign its first place. It could be that this criterion may have some range of variation that does not affect the best alternative, or it could be that the range is very small or even that there is no range.If the DM, consulting statistics and other sources, finds that the pair had considerable variation in a year, for instance, then he can decide that D, even being the best alternative selected by a model, is not the best, and he selects the second best. It is important to consider that the DM must have credible reasons to perform this change.Sensitivity analysis, in most MCDM methods, uses the weights of criteria, pre-determined for instance, using AHP, and then selects for it the criteria with its highest weight, while maintaining the weights of the other criteria constant, and observing the performance of the selected alternative. This is usually done using a graphic.

Simulation   Mathematical process is consistent in preparing a mathematical model of the system to analyse and running it hundreds or even thousands of times using random values for the variables.

SIMUS (Munier 2011)   (Sequential Interactive Model for Urban Systems). It is a hybrid MCDM method mainly based in Linear Programming but also incorporating Simple Additive Weighting and Outranking.

Stocking   Keeping a stock of something, for instance raw materials, spare parts, water and so on. Sometimes, the scenario calls for using the stocks but keeping, for diverse reasons, a minimum amount. In other occasions, the scenario may call for using stock up to a certain limit, that cannot be surpassed and also it can call for using the whole stock.

Sorting　Ordering following an increasing sequence, according to user preferences, and keeping consistency, for instance, 1, 2, 3, 4, where the user wants it in an increasing sense. It is also the ordering of alternatives into a homogenous group.

Stakeholder　Interested party of the project.

Subjective weights　Weights determined based on the DM preferences.

System　According to Law and Kelton (1991), *A system is a collection on entities that act and interrelate together toward the accomplishment of some logical end.* It is important here the mention that in a system the element interrelate together.

TOPSIS method (Hwang and Yoon 1981)　An MCDM method based on the distance to an ideal solution.

Trade-off values　According to the Dictionary, it can be defined as *Giving up of one thing in return for another, or sacrificing something to get a gain in another.*

Transitivity　In MCDM, if we have a sequence of events such as A, B and C, then there is transitivity when there is agreement in a comparison. For instance, if A is larger than B, and B is larger than C, then there is an understanding that A is larger than C. Transitivity, a concept of mathematical logic, is not often found in the real world. For instance, a person may compare her vacation in three different potential spots such as relaxing at home or going to a mountain quiet resort or enjoying active social life on a beach. She may prefer staying at home to going to the mountain and preferring the mountain to the beach, but it does not necessarily mean that she will prefer staying at home rather than going to the beach.

Verbal-Numerical comparison　The act of assigning a cardinal to a linguistic comparison.

Weighted sum　The sum of the products between the performance in a criterion times the criterion weight.

References

Alonso, J., & Lamata, M.-T. (2006). Consistency in the analytic hierarchy process: A new approach. *International Journal of Uncertainty, Fuzziness and Knowledge-Based Systems, 14*(4), 445–459.

Arrow, K. (1951). *Social choice and individual values.* Doctoral thesis (1st edn). New York: Wiley.

Asadabadi, M., Chang, E., & Sabari, M. (2018). Are MCDM methods useful? A critical review of analytic hierarchy process (AHP) and analytic network process (ANP). *Cogent Engineering, 6*(1), 1.

Bana e Costa, C., & Vansnick, J.-C. (2008, June 16). A critical analysis of the eigenvalue method used to derive priorities in AHP. *European Journal of Operational Research, 187*(3), 1422–1428.

Barakos, G. (2019). *Communication* in ResearchGate.

Barba-Romero, S. (2000). *El análisis de la toma de decisiones.* https://studylib.es/doc/231559/evaluacion-multicriteria-introduccion

Barzilai, J. (2001). *Notes on the analytic hierarchy process* – ResearchGate.

Belton, V., & Gear, T. (1983). On a shortcoming on Saaty's Method of analytic hierarchies. *Omega, 11*(1983), 228–230.

Belton, V., & Stewart, T. (2002). *Multiple criteria decision analysis: An integrated approach.* Boston: Kluwer Academic Publishers.

Belton, V., & Stewart, T. (2010). Problem structuring – Final preprint of paper published as Chapter 8, pp. 209–239. In M. Ehrgott, J. R. Figueira & S. Greco (Eds.), *Trends in multiple criteria decision analysis.* Springer.

Bernasconi, M., Choirat, C., & Ser, R. (2007). *The analytical hierarchy process and the theory of measurement.* Report Universitat Ca Foscari, Universitat dell'Insubria, Universite of Navarra.

Bozóki, S., & Rapcsák, T. (2008). *On Saaty and Koczkodaj's inconsistencies of pair-wise comparison matrices.* Hungarian Academy of Sciences.

Brans, J., & Vincke, P. (1985, June). A preference ranking organisation method: (The PROMETHEE method for multiple criteria decision-making). *Management Science, 31*(6), 647–656.

Buchanan, J., Henig, E., & Henig, M. (1998). Objectivity and subjectivity in the decision-making process. *Annals of Operations Research, 80*(1998), 333–334.

Carlsson, C., Fuller, R., & Björk, K-M (2008). *Problem solving with multiple interdependent objectives.* ResearchGate.

Checkland, P. (1983). O.R. and the systems movement: Mappings and conflicts. *Journal of the Operational Research Society, 34*(8), 661–675. https://doi.org/10.1057/jors.1983.160.

Chung, S., Lee, H., & Pearn, W. (2005). Analytical network process (ANP), approach for mix planning in semiconductor fabricator. *International Journal of Production Economics, 96*, 15–36.

Churchman, C., Ackoff, R., & Arnoff, E. (1957). *Introduction to operations research* (p. 1957). New York: Wiley.

Cilliers, P. (2001, June). Boundaries, hierarchies and networks in complex systems. *International Journal of Innovation Management, 5*(2), 135–147.

Cinelli, M., Coles, R., & Kirwan, K. (2014). Analysis of the potentials of multi criteria decision analysis methods to conduct sustainability assessment. *Ecological Indicators, 4*, 138–148.

Cohon, J., Facet, T., Haan, A., & Marks. D. (1973). *Mathematical programming models and methodological approaches for river basin planning.* Technical report. Cambridge, MA: Ralph M. Parsons Lab for Water Resources and Hydrodynamics, MIT.

Cole, M. (2018). *Is this why you can't solve complex problems?* Medium/Leadership.

Dantzig, G. (1948). *Linear programming and extensions.* R-366-PR- Corporation.

Department of Transport, Local Government and the Regions (DTLR) (UK Government). (2001).

Dodd, F., Donegan, H., & McMaster, T. B. M. (1995). Inverse inconsistency in analytic hierarchies. *European Journal of Operational Research, 80*(1), 86–93.

Dyer, J. (1990a). A Clarification of remarks on the analytic hierarchy process. *Management Science, 36*(3), 274–275.

Dyer, J. (1990b). Remark on the analytic hierarchy process. *Management Science, 36*(1990), 249–258.

Dyer, J., & Wendell, R. (1985). *A critique of the analytic hierarchy process.* Working paper 84/85–424. Department of Management, The University of Texas at Austin.

Farkas, A. (2007). The analysis of the principal eigenvector of pairwise comparison matrices. *Acta Polytechnica Hungarica, 4*(2), 99–115.

Fishburn, P. C. (1967). *Additive utilities with incomplete product set: Applications to priorities and assignments.* Baltimore, MD: Operations Research Society of America (ORSA).

Fishburn, P. (1991, April). Nontransitive preferences in decision theory. *Journal of Risk and Uncertainty, 4*(2), 113–134.

Forman, E. H. (1990). AHP is intended for more than expected value calculations. *Decision Sciences, 21*(3), 670–672.

Forman, E. H. (1993). Facts and fictions about the analytic hierarchy process. *Mathematical and Computer Modelling, 17*(4–5), 19–26.

Franek, J., & Kresta, A. (2014). Judgement scales and consistent measure in AHP. *Procedia Economics and Finance, 2014*(12), 164–173.

French, S. (1983). *Decision theory: An introduction to the mathematics of rationality.* Chichester: Ellis Horwood.

García-Cascales, M., & Lamata, M. T. (2011). On rank reversal and TOPSIS method. *Mathematical and Computer Modeling, 56*(2012), 123–132.

Goepel, K. (2019). *Simple Priority Calculator.* https://bpmsg.com/ahp/ahp-calc.php.

Grupp, A., & Schubert, T. (2010). *Review and new evidence on composite innovation indicators for evaluating national performance.*

Habenicht, W., Scheubrein, B., & Scheubrein, R. (2002). Multiple-criteria decision making. *Optimization and Operations Research, IV.*

Harker, P., & Vargas, L. (1990). Reply to 'remarks on the analytic hierarchy process'. *Management Science, 36*(1990), 269–273.

Hashemkhani, S., Maknoon, R., Zavadskas, E. K. (2016). Multiple Attribute Decision Making (MADM) based scenario. *International Journal of Strategic Property Management, 20*(1), 101–111. ISSN 1648-715x. eISSN 1648-9179.

Hazelrigg, G. (2019). *Communication* in ResearchGate.

Hobbs, B., Bell, M., & Ellis, H. (2003, December). The use of multi-criteria decision-making methods in the integrated assessment of climate change: Implications for IA practitioners. *Socio-Economic Planning Sciences, 37*(4), 289–316.

Hulkower, N., & Neatrour, N. (2016). Deflecting arrow by aggregating rankings of multiple correlated criteria according to Borda. *Journal of Multi-Criteria Decision Analysis, 23*, 75–86.

Hwang, C., & Yoon, K. (1981). *Multiple attribute decision making: Methods and applications.* New York: Springer.

ICDSST. (2015). *decision support systems v – Big data analytics for decision making* (pp 98–109).

Ihimekpen, N., & Isagba, E. (2017, March). The use of AHP (analytical hierarchy process) as multi criteria decision tool for the selection of best water supply source for Benin City. *Nigerian Journal of Environmental Sciences and Technology (NIJEST), 1*(1), 169–176.

Improta, G., Russo, M., Trassi, M.-T., Converso, G., Murino, T., & Santillo, L. (2018). *Use of the AHP methodology in system dynamics: Modelling and simulation for health technology assessments to determine the correct prosthesis choice for hernia diseases Giovanni.* Naples: Department of Public Health, University of Naples "Federico II".

International Journal of Innovation Management, *5*, 2 (2001, June), 135–147. Imperial College Press.

Ishizaka, A, & Labib, A. (2009). *Analytic Hierarchy Process and Expert Choice: Benefits and limitations.*

Ishizaka, A., & Labib, A. (2011). Review of the main developments in the analytic hierarchy process. *Expert Systems with Applications, 38*(11), 14336–11434.

Ishizaka, A., & Siraj, S. (2018, January 16). Are multi-criteria decision-making tools useful? An experimental comparative study of three methods. *European Journal of Operational Research, 264*(2), 462–447.

Jiménez Moreno, J.-M., Casas Altuzarra, A., Urmeneta, M., & Escobar, T. (2004). *El índice de consistencia geométrico para matrices incompletas en AHP.* Grupo Decisión Multicriterio – Universidad de Zaragoza.

Kahnemann, D., & Tversky, D. (1974, September 27). Judgment under uncertainty: Heuristics and biases. *Science New Series, 185*(4157), 1124–1131.

Kantorovich, L. (1939). Mathematical methods of organizing and planning production. *Management Science, 6*(4), 366–422.

Kasperczyk, N., & Knickel, K. (2006). *The Analytic Hierarchy Process (AHP).* Available at: www.ivm.vu.nl/en/Images/MCA3_tcm53-161529.pdf.

Kazibudzki, P. (2013). On some discoveries in the field of scientific methods for management within the concept of analytic hierarchy process. *International Journal of Business and Management, 8*(8).

Keeney, R. (1992). *Value-focused thinking: A path to creative decision making.* Cambridge, MA: Harvard University Press.

Kenney, R., & Raiffa, H. (1976). *Decisions with multiple objectives: Preferences and value trade-offs.* New York: Wiley.

Kimball, Z. (1957, June). Errors of the third kind in statistical consulting. *Journal of the American Statistical Association, 52*(278), 133–142.

Koen, R., (2008). *Aspects of MCDA classification and sorting methods.* Master thesis. University of South Africa (UNISA).

Köksalan, M., Wallenius, J., & Zionts, S. (2013). An early history of multiple criteria decision making. *Journal of Multi-Criteria Decision Analysis, 20*, 87–94.

Konidari, P., & Mavrakis, D. (2007, December). A multi-criteria evaluation method for climate change mitigation policy instruments. *Energy Policy, 35*(12), 6235–6257.

Ksenija, M., Bobar, V., & Delibašić, B. (2015). *Modeling interactions among criteria in MCDM methods: A review.*

Kunsch, P. L. (2012). *Why pairwise comparison methods may fail in MCDM rankings.* Vrije Universiteit Brussel.

Law, A., & Kelton, W. (1991). *Simulation, Modeling and Analysis.* McGraw-Hill.

Lee, M.-C. (2010). The analytic hierarchy and the network process in multicriteria decision making: performance evaluation and selecting key performance indicators based on ANP Mode. In *Convergence and hybrid information technologies.* https://doi.org/10.5772/9643.

Leung, L.,, & Cao, D. (2001, July 1). On the efficacy of modeling multi-attribute decision problems using AHP and Sinarchy. *European Journal of Operational Research, 132*(1), 39–49.

Lin (2008). Teng (2011), Xu (2008), Yao (2010). In *An introduction to prospective multiple attribute decision.* Zolfani, Segadhat, Maknoon, Zavadskas. *Technological and Economic Development of Economy, 22*(2), 2016.

Liu, H. S., Yeh, Y., & Huang, J. (2014). Correlated analytic hierarchy process. *Mathematical Problems in Engineering 2014*, 961714, 7 p.

Ma, J., Fan, Z.-P., & Huang, L. (1999, January 16). A subjective and objective integrated approach to determine attribute weights. *European Journal of Operational Research, 112*(2), 397–404.

Massachusetts Institute of Technology (MIT). *Report.*

Moshkovich, H., Gomes, L. F. A. M., Mechitov, A. I., & Rangel, L. (2012). Influence of models and scales on the ranking of multi attribute alternatives. *Pesquisa Operacional, 32*(3), 523–542.

Munier, N. (2011). *Procedimiento fundamentado en la programación lineal para la selección de alternativas en proyectos de naturaleza compleja y con objetivos múltiples.* PhD thesis. Universidad Politécnica de Valencia, España.

Munier, N., Hontoria, E., & Jimenez, F. (2019). *Strategic approach in multi-criteria decision making – A practical guide for complex scenarios* (p. 2019). New York: Springer.

Opricovic, S., & Tzeng, G. (2004). Compromise solution by MCDM methods: A comparative analysis of VIKOR and TOPSIS. *European Journal of Operational Research, 156*(2), 445–455.

Öztürk, Z. (2006). *A review of multi criteria decision making with dependency between criteria.* Turkey: Anadolu University.

Pérez, J. (1995). Some comments on Saaty's AHP. *Home Management Science, 41*(6).

Pérez, J., Jimeno, J., & Mokotoff, F. (2006). Another potential shortcoming of AHP. *TOP, 14*(1), 99–111.

Peters, M., & Zelewski, S. (2008). *Pitfalls in the applications of analytic hierarchy process to performance measurement.* Essen: Institute for Production and Industrial Information Management/University of Duisburg-Essen.

Ramík, J., & Perzina, R. (2014). Solving decision problems with dependent criteria by new fuzzy multicriteria method in excel. *Journal of Business & Management, 3*(4), 1–16.

Rodrigues, T., Montibeler, G., Olivera, M., & Bana e Costa, C. (2017, May 1). Modelling multi-criteria value interactions with reasoning maps. *European Journal of Operational Research, 258*(3), 1054–1071.

Roy, B. (1968). La méthode ELECTRE. *Revue d'Informatique et de Recherche Opérationnelle (RIRO), 8*, 57–75.

Saaty, T. (1980). *The analytic hierarchy process.* New York/Pittsburgh: McGraw Hill/RWS Publications.

Saaty, T. (1986). Axiomatic Foundation of the Analytic Hierarchy Process. *Management Science, 32*(7), 841–855.

Saaty, T. (1990). An exposition of the AHP in reply to the paper "remarks on the analytic hierarchy process". *Management Sciences, 36*(3), 259–268.

Saaty, T. (1994). *Fundamentals of decision making and priority theory with the analytic hierarchy process* (Vol. 6). Pittsburgh: RWS Publications.

Saaty, T. L. (1996). *Decision making with dependence and feedback: The analytic network process.* Pittsburgh: RWS Publications.

Saaty, T. (1997). That is not the analytic hierarchy process: What the AHP is and what it is not. *Journal of Multi-Criteria Decision Analysis, 6*(6), 324–335.

Saaty, T. (2001). *Decision-making with the AHP: Why is the principal eigenvector necessary.* ISAHP 2001, Berne, Switzerland.

Saaty, T. (2008a). Decision making with the analytic hierarchy process. *International Journal of Services Sciences, 1*(1), 83.

Saaty, T. (2008b). The analytical network process. *Iranian Journal of Operations Research, 1*(1).

Saaty. R. (2016). *Book decision making in complex environments- The Analytic Network Process (ANP) for dependence and feedback.* Super Decisions- Software for Decision Making with Dependence and Feedback.

Saaty, T., & Vargas, L. (1991). *Prediction, projection and forecasting.* Boston: Kluwer Academic.

Salo, A., & Hämäläinen, R. (1992). *On the measurement of preferences in the analytic hierarchy process,* Research Report. A47, Helsinki University of Technology, Systems Analysis Laboratory, Espoo, Finland.

Salo, A., & Hämäläinen, R. (1997). On the measurement of preferences in the analytic hierarchy process. *Journal of Multi-Criteria Decision Analysis, 6*, 309–319.

Salvia, A., Brandli, L., Filho, C., & Locatelli Kalil, R.-M. (2019). An analysis of the applications of Analytic Hierarchy Process (AHP) for selection of energy efficiency practices in public lighting in a sample of Brazilian cities. *Energy Policy, 132*(2019), 854–864.

Schoner, B., & Wedley, W. C. (1989). Ambiguous criteria weights in AHP: Consequences and solutions. *Decision Sciences, 20*(3), 462–475.

Sehra, S., Singh Bar, Y., & Kaur, N. (2012, January). Multi criteria decision making approach for selecting effort estimation model. *International Journal of Computer Applications* (0975–8887), *39*(1).

Shannon, C. (1948). A mathematical theory of communication. *Bell System Technical Journal.*

Smith, J., & Winterfeldt, D. (2004). Decision analysis in management science. *Management Science, 50,* 561–557.

Song, B., & Kang, S. (2016). A method of assigning weights using a ranking and non hierarchy comparison. *Advances in Decision Sciences, 2016,* 8963214, 9 p.

Stanujkic, D., Đorđević, B., & Đorđević, M. (2013). Comparative analysis of some prominent MCDM methods: A case of ranking Serbian banks. *Serbian Journal of Management, 8*(2), 213–241.

Thurstone, L. (1927). A law of comparative judgment. *Psychology Review, 34*(1927), 273–286.

Tomashevskii, I. (2015, February 1). Eigenvector ranking method as a measuring tool: Formulas for errors. *European Journal of Operational Research, 240*(3), 774–780.

Triantaphyllou, E. (2000). *Multi-criteria decision-making methods: A comparative study.* Cham: Springer.

Triantaphyllou, E. (2001). Two new cases of rank reversals when the AHP and some of its additive variants are used that do not occur with the multiplicative AHP. *Journal of Multi-Criteria Decision Analysis, 10*(1), 11–12.

Triantaphyllou, E., & Mann, S. (1989). An examination of the effectiveness of multidimensional decision-making methods: A decision-making paradox. *International Journal of Decision Support Systems, 5,* 303–312.

Triantaphyllou, E., & Mann S. (1994, July). A computational evaluation of the revised analytic hierarchy process. *Computers & Industrial Engineering.*

Triantaphyllou, E., Shu, B., Nieto Sanchez, S., & Ray, T. (1998). Multi-criteria decision making: An operations research approach. In J. G. Webster (Ed.), *Encyclopedia of electrical and electronics engineering* (Vol. 15, pp. 175–186). New York: Wiley.

Uniersia. (2009). *Net of Latin America and Spain universities.*

Vargas, L. (1982). Reciprocal matrices with random coefficients. *Mathematical Modeling, 3,* 69–81.

Velasquez, M., & Hester, P. (2013). An analysis of multi-criteria decision-making methods. *International Journal of Operations Research, 10*(2), 56–66.

Velten, K. (2009). *Mathematical modeling and simulation- introduction for scientists and engineers.* Weinheim: Wiley-VCH.

Vujičić, M., Papic, M., & Blagojević, M. (2016) *Comparative analysis of objective techniques for criteria weighing in two MCDM methods on example of an air conditioner selection.* https://doi.org/10.5937/tehnika17034.

Wang, X., & Triantaphyllou, E. (2008). Ranking irregularities when evaluating alternatives by using some ELECTRE methods. *Omega, 36*(1), 45–63.

Wang, C., & Yoon, K. (1981). *Multiple attributes decision making methods and applications.* Berlin: Springer.

Warren, L. (2003). *Uncertainties in the analytic hierarchy process.* Report.

Watson, S., & Freeling, A. (1983). Comment on: Assessing attribute weights by ratios. *Omega, 11,* 13.

Weber, M. (1997). Remarks on the paper: On the measurement of preferences in the analytic hierarchy process. *Journal of Multicriteria Decision Analysis, 6,* 320–321.

XLSTAT (NA). (2019). *Basic, essential data analysis tools for Excel-* https://help.xlstat.com/customer/en/portal/articles/2961443-analytic-hierarchy-process.

Yeh, C. H. (2002). A problem-based selection of multi-attribute decision-making methods. *International Transactions in Operational Research, 9*(2), 169–181. https://doi.org/10.1111/1475-3995.00348.

Yüksel, I., & Deviren, M. (2007, August 15). Using the analytic network process (ANP) in a SWOT analysis – A case study for a textile firm. *Information Sciences, 177*(16), 3364–3382.

Zadeh, L. (1965). Fuzzy sets. *Information and Control, 8*(3), 338–353. https://doi.org/10.1016/S0019-9958(65)90241-X.

Zanakis, S. H., Wishart, N., & Durlish, S. (1998). Multi-attribute decision making: A simulation comparison of select methods. *European Journal of Operational Research, 107*(3), 507–529.

Zardari, N., Ahmed, H., Shirazi, K., & Yusop, Z. (2015). *Weighting methods and their effects on multi-criteria decision-making model outcomes in water resources management*. Cham: Springer.

Zeleny, M. (1974, December). A concept of compromise solutions and the method of the displaced ideal. *Computers & Operations Research, 1*(3–4), 479–496.

Zeleny, M. (2011). Multiple criteria decision making (MCDM): From paradigm lost to paradigm regained? *Journal of Multi-Criteria Decision Analysis, 18*, 77–89.

Index